SHEEP HUNTING IN ALASKA

The Dall Sheep Hunter's Guide

by TONY RUSS

Copyright© 1994 by Tony Russ

All rights reserved. No part of this book may be used or reproduced in any form or by any means without prior written permission of the Publisher, except in the case of brief quotations as part of critical reviews and articles.

First Printing 1994

Cover photograph by Tom Soucek© 1993
Prints available from Tom Soucek Photography
907-248-4164

Library of Congress Catalog Card Number: 93-86886
ISBN 0-9639869-0-2

Northern Publishing
574 Sarah's Way
Wasilla, AK 99654

CONTENTS

Acknowledgments	8
Introduction	10

PART I–PREPARATION

Chapter 1... **WHY HUNT SHEEP**   13

Chapter 2... **GETTING INTO SHEEP SHAPE**   19
         Cardiorespiratory Conditioning.
         Strength. Flexibility and Agility.
         Stamina. You are What you Eat.

Chapter 3... **FOOTWEAR AND CLOTHING**   31
         Footwear. Socks. Clothing. Rainwear.

Chapter 4... **BACKPACKING GEAR**   47
         Backpacks. Tents. Sleeping Bags.
         Stoves. Cookware. Water Containers.
         Climbing Poles. First Aid. Etc.
         Organization. Camouflage.

Chapter 5... **WEAPONS AND OPTICS**   63
         Firearms. Archery Equipment.
         Optics.

Chapter 6... **FOOD FOR THE HIGH COUNTRY**   71
         Nutritional Needs. Calories per Ounce.
         Freeze-Dried Foods. Pasta.
         Ready-to-Eat Foods. Drinks.
         Wild Foods. Supplements. Packaging.

PART II–HUNTING

Chapter 7... **SHEEP BEHAVIOR**   85
        Population Dynamics. Social Behavior Patterns. Defenses. Feeding Habits. Resting Habits. Seasonal Movements. Lifetime Habits, Movements.

Chapter 8... **HUNTING STRATEGIES**   101
        Basics for the Hunter. Basics of the Hunt. The Approach–From Above or Below? Trojan Sheep. Bowhunting. Timing of the Hunt. Transportation.

Chapter 9... **CARE OF MEAT AND TROPHIES**   119
        Skinning. Meat and Horns. Quality Care. Preserving the Trophy.

Chapter 10.. **A TROPHY SHEEP**   125
        Legal Rams. Record-Book Rams. Field-Judging.

Chapter 11.. **WHERE TO HUNT**   135
        Information Sources. FNAWS. Dall Sheep Management in Alaska. Brooks Range. Alaska Range. Tanana/Yukon Uplands. Kenai Mts. Talkeetna Mts. Chugach Mts. Wrangell Mts. What do **YOU** Want?

Chapter 12.. **THOUGHTS ABOUT HUNTING**   147

Appendix...   151
Boone and Crockett Score Sheet   152
Safari Club International Score Sheet   155
References   157
Gear Checklist   159

ACKNOWLEDGMENTS

There are many people I need to thank for helping me with this book. I first want to thank Joan L. Baxter, my high school English teacher, who still writes to me. She was an excellent instructor and still inspires me.

Thanks to Lynn Soiseth who always shares information about large rams and Bill Stevenson who showed me where to find them. Thanks to Bob Hodson for sharing his knowledge about gear and Carl Brent for his assistance with several aspects of this book.

Wayne E. Heimer and David Harkness are the state biologists largely responsible for statistics and scientific papers about Dall sheep in Alaska. These sources were invaluable to me in supporting this text. As both a writer and sheep hunter, thank you.

I want to credit Bill Parker for getting me into bowhunting. I've learned more as a bowhunter than ever imaginable as a rifle hunter.

I must give credit to Dave and Kris Widby for their photographic and editorial help. And Jay Massey for getting me started in the right direction when I began this project.

Thanks to Diane O'Loughlin and Frederick L. Wedel who performed the final proofreading.

Most of all I want to thank Lon E. Lauber, my editor. His expertise was invaluable to me. His background in hunting and journalism is extensive; I couldn't have done **half** as well without him—and that is probably an **understatement.**

Also, thanks to all those who I have not mentioned. Your help during the past 18 months is truly appreciated.

Lon E. Lauber, the book's Editor, with one of his Pope and Young rams. As this photo illustrates, Lon has first-hand knowledge of the physical demands and adverse weather conditions sheep hunters have to overcome. Lon is a free-lance outdoor writer/photographer and an accomplished bowhunter. 37 3/8 x 12 6/8. Score—154 7/8 P&Y.

INTRODUCTION

I first hunted Dall sheep in 1978 and am **fascinated** by them. I like watching, hunting, and photographing them. I enjoy talking about them and traveling through the country they live in. I've learned a considerable amount about Dall sheep and how to hunt them. I continue to learn every time I encounter them. Writing this book has taught me even more. I look forward to talking to people with more sheep hunting experience than myself as well as less--experienced hunters. **Everyone** has something to teach me.

Dall sheep hunting in Alaska is almost always a backpacking experience. My sheep hunting addiction has taught me how to safely and comfortably travel in the mountains sheep inhabit. This includes information about conditioning, footwear, clothing, gear selection, and suitable backpacking food. This is necessary information for hunting Dall sheep, but it is also useful when hunting **any** animal or pursuing **many other** outdoor activities in Alaska.

This guide is organized with the first six chapters devoted to preparation--which is at least **half** of any hunt. There is one chapter on sheep behavior and one chapter on hunting stategies. The next three chapters are basically more information necessary to plan and complete a hunt. The last chapter is simply some extra thoughts about hunting in general. Only two chapters out of twelve deal directly with sheep hunting. This is **appropriate** because preparation and planning are the major components of any hunt.

Personally, I bowhunt for sheep. However, there are less than ten pages in this book dealing **directly** with bowhunting. This is **also** appropriate because 90% of planning, preparation, and hunting is the same for gun hunters and bowhunters.

I wrote this book to help others hunt Dall sheep--Ovis dalli--which inhabit Alaska and Canada. Most of the Dall sheep hunting in Alaska is open to anyone with a general hunting license. There are only a relatively few **drawing** permit areas. So residents--like myself--have the opportunity to hunt this magnificent animal **every** year. Since nonresidents of Alaska or Canada are required to hire a guide to hunt Dall sheep, the cost prohibits most from acquiring much experience. Logically, a resident of Alaska or Canada would write a book on Dall sheep hunting to pass along their knowledge. However, noone **has** previously written a book on Dall sheep hunting! Hopefully, this book will fill that need.

PART ONE- PREPARATION

Chapter 1

WHY HUNT SHEEP

The Dall ram was totally involved in his feeding. I peered down at him from a rocky spine less than 50 yards away—waiting for the best opportunity to shoot. During the past 30 minutes his feeding had brought him in from 150 yards to within bow range.

He appeared confident in his location. He never looked up for danger. In fact, **I** shouldn't have clawed my way up the cliffs after him. However, like too many sheep hunters, I have more drive than caution.

When the ram turned his back toward me, I stood up on the ridge and sent an arrow on it's way. The hit was good. The ram traveled downhill immediately. I slowly followed and found him bedded, near death. I put another arrow into him at close range. I didn't want to risk him getting up and then falling down the 2,500 vertical feet of steep cliffs and slides.

My second shot, although good, was a mistake. The ram stood up, stumbled a few steps, and died. He was only 50 yards away, but I couldn't catch up before he started to roll. The dead ram picked up speed. I watched in horror as he slid, bounced, and free-fell seven hundred feet before stopping. Agonizing over his possible condition, I **carefully** climbed down to the ram.

The ram lay precariously at the bottom of a steep slope, just above a cliff. I held onto the rock face with one hand while I skinned, boned, and loaded him with the other. When I finished, daylight was fading and it had started to rain.

I considered going back up, over, and down the way I came. However, I knew darkness would catch me on a cliff, with little water, in the rain. I decided to go down.

I had about 1,500 feet of unfamiliar rock to negotiate with a loaded pack. I was very apprehensive, but felt it was my best option.

Ambition often overcomes prudence. As a result, sheep hunters frequently find themselves descending with a heavy load after dark.

Picking the route was very difficult because I could only see the closest 100 feet below me. The curve of the slope hid the rest of the mountain from view.

WHY HUNT SHEEP

I meticulously picked my way down to avoid a fatal slip. I had to descend several 70° faces, clambering like a cliff monkey. I was only 200 feet short of the bottom at dusk. I was just beginning to think I had made it when my route disappeared. Broken cliffs and slides below me offered no path unless I descended 85° to 90° cliffs.

My anxiety level soared. I considered throwing my pack over a cliff to make my descent less risky, but then I would be without **any** gear if I didn't get down. I had to go for it with my pack.

The 20 feet of rope I carried was just long enough to lower my pack down the shortest cliff. Luckily, the pack didn't roll away when I let go of the rope. I descended using two-inch ledges for handholds. I made it. Then, there was only one more frightening ten-foot drop before the bottom.

I reached my tent ten minutes before dark and rejoiced. When I noticed a bear had slashed my tent and there was rainwater inside, I didn't much care. I was just relieved to be alive.

Events like these occur frequently while sheep hunting. It is sometimes hard to justify our reasons for hunting sheep; especially to someone who is not hooked, like many of us are. However, some of the reasons to hunt sheep are understandable.

One basic reason we hunt is to provide food for ourselves, families, and friends. Regardless of how sophisticated our world becomes and abstract our occupations seem, we **still** have a basic drive to provide. Hunting gives us an opportunity to feel satisfaction in our physical abilities—a rare thing in our increasingly sedentary society. Also, wild sheep meat is low-fat, high-protein, unmatched table fare.

A Dall sheep trophy is magnificent. This is reason enough for many of us to continue hunting them. With their white coats and dark, curling horns, they stand out in any trophy display. Some sheep populations in Alaska tend to produce larger horns than others. And some populations have other horn characteristics which make them more-desirable trophies. However, any Dall ram is a trophy the hunter should be proud to display.

The physical challenge compels many of us to hunt these white sheep. The physical demands of sheep hunting are extreme. The quantity and quality of physical conditioning a hunter invests in the hunt often determine the outcome. A successful Dall sheep hunt is recognized as a significant accomplishment—and justifiably so.

This photo of King Cavalier and his ram illustrates two reasons why we hunt sheep. Dall sheep are magnificent animals and they live in spectacular country. 36 x 13.

Sheep hunters must also meet a mental challenge. They must **commit** to sufficient preparation as well as the demands of an actual hunt. Hunters should plan on being cold, wet, tired, hungry, thirsty, and sore—maybe all at the same time! The rigors of a sheep hunt have beaten many hunters. Mental preparation and toughness are crucial to success.

One less-tangible reason we hunt sheep is to escape. During a hunt, we can focus our entire being on one objective—and put aside the worries of our everyday lives. We often return from a hunt with renewed enthusiasm, appreciation, and satisfaction with our lives.

There is also adventure, beautiful scenery, and the magnificence of the animals themselves. Every hunter will have a unique mix of reasons to hunt Dall sheep. The reasons I listed have only partially explained the fascination we have for hunting them.

A casual conversation I had during a recent medical appointment may help illustrate the allure of hunting these animals. While drawing my blood, a female staff member recalled her first sheep hunt. This young woman had been coaxed into accompanying her fiance' on a hunt. Halfway up the mountain she questioned the sanity of

WHY HUNT SHEEP

continuing. At this critical moment, they spotted a full-curl ram silhouetted on the skyline above. She said that's all it took. She never questioned why again and is anxiously waiting for her next sheep hunt.

It is difficult to explain why we hunt sheep. A hunter who is wet, tired, foot-sore, bug-bitten, and hours away from the top frequently asks himself "why?"

Your reasons for hunting sheep will affect your approach to the next phase of preparation—**physical conditioning**.

A view of a skylined ram may best communicate why many of us hunt Dall sheep.

Chapter 2

CONDITIONING

Dall sheep inhabit steep, rough terrain in remote country. Hunting them is always a very physical adventure. The mountains are steep, the distances great, and the packs heavy. The success of any sheep hunt will depend to a great extent on the hunter's physical condition. Each hunter must consider the questions raised in the first chapter and find their personal reasons to hunt sheep. These reasons will determine how much effort they put into physical conditioning.

Most of us should exercise to prepare for a sheep hunt. There are a few lucky hunters whose occupations provide sufficient physical exercise to prepare them for the hunt. However, if you are not one of those select few, you need a conditioning program. But remember to get a medical okay from your doctor before beginning any new exercise program.

The suggestions I am about to make concerning physical conditioning are just that—only suggestions. Each hunter should design an exercise program that takes into account his/her **current** health, time constraints, and ambitions. The more accurately you analyze **your** situation, the more effectively you can prepare for a sheep hunt (or anything in life).

I like to use a cross-training approach to physical conditioning. I have divided my program into four parts:
-cardiorespiratory conditioning
-strength training
-flexibility and agility training
-stamina

Each hunter needs to evaluate him- or herself with respect to these four parameters. Train **your** weaknesses. Realize the limitations a weak back or poor wind will put on you. Some weaknesses can be realistically improved; others cannot. Make an accurate assessment and then design a program that most efficiently improves your situation.

I remember one guided sheep hunt I went on that vividly illustrates the need for a cross-training approach to get into top "sheep shape". The hunter I was guiding was a marathon runner and in excellent shape—for running. He did not have any excess bodyfat and his cardiorespiratory conditioning was great. On the trip in he carried 25 lbs. I needed to stop and catch my breath more often than him. This was my first hunt of the season and my cardiorespiratory system wasn't in great shape—yet. But on the way out, when I packed 95 lbs. to his 55 lbs., he was dragging long before we made the airstrip. His final words as he set down his pack were "If Bill [the pilot] doesn't come in, my bones will be here forever, because I'm not carrying that pack one more step." A sheep hunt requires physical fitness in all four of the areas I have listed. Any shortcomings in a hunter's physical conditioning will show up during a hunt.

Cardiorespiratory Conditioning

Climbing steep slopes at high altitudes with a heavy pack requires good cardiorespiratory (c/r.) fitness. Aerobic exercise improves your c/r. fitness and can be any activity that makes you breathe hard for a long time. Running, biking, swimming, racquetball, and of course mountain climbing are some of the ways to improve your c/r. fitness. Personally, I like the stationary bike, rowing machine, and the stepper/climbers that have become so popular lately. The stepping machines can duplicate the climbing motion that all sheep hunters learn to love(?). With one of these machines, I can still exercise my legs and maintain my c/r. conditioning during the winter months. Even in the summer when I don't have time to train on a nearby mountain, I can still improve my climbing ability and prepare for the upcoming season. I also use a stationary bike and a rowing machine (I own both) for variety in my aerobic training. I find that by varying the exercises I experience considerably fewer problems with my joints. And, I don't get into a rut as easily if I use a variety of machines.

CONDITIONING

For those hunters who are new to sheep hunting, you should consider getting some climbing experience before the hunt. The mountains Dall sheep inhabit are steep and dangerous. Sheep hunters often climb fifty- or sixty-degree slopes. This requires use of their hands as well as feet. This can be **frightening** for someone who is afraid of heights. I advise you to do some mountain climbing if you have no experience and doubt your ability to climb steep, rocky faces. It is great exercise for your body and will also help prepare you mentally for the hunt.

Good aerobic conditioning is required to climb 30° slopes like this one the author is contemplating.

The standard pattern for aerobic training is to work out at least 3 times per week, at least 20 minutes each time, at your target heart rate. Your target heart rate is determined by subtracting your age from 220 and multiplying the result by 80%(.80). A 40-year-old would have a target rate of 144 beats per minute[(220-40) x .80]. So whether you like to run, swim, or use apparatus like I do, you can use this guideline to begin with. Of course, if you want to get into better than just passable shape, and your doctor has given you the okay, you will need to do more than this basic aerobic training. Your goals and the hunt you have planned will determine your training program.

Strength

Brute strength can be very useful on a backpack hunt—as experienced sheep hunters can verify. Strength not only helps you overcome difficult physical obstacles during a hunt, but it also allows you to continue at a sub-maximum level for a much longer time. It is much easier to carry a 100 lb. pack all day if you have regularly trained with a 120 lb. pack. Most of our strength lies in the large muscles of our legs, hips, and lower back. These are the muscles you should train for strength.

Leg, hip, and back muscles can be trained for strength most effectively with weights. Leg presses, squats, leg extensions, lunges, and calf raises are standard exercises for legs and hips. Good back exercises include deadlifts and good mornings. There are numerous modifications of these basic exercises that can be found in any weight-training guide or by asking any competent weight-room attendant. In general, I use moderate repetitions (8-15) when I am training for strength. If you are not familiar with weight training, you should first consult a good trainer. Weight training can dramatically increase your lower body strength if it is done properly.

Another good way to train your legs is to climb stairs. Long staircases are available to almost everyone and they are free to use. Make climbing stairs a lunchtime activity if you work near or in a tall office building. You can even make this closer to the real thing by carrying a pack. This may draw strange looks from others, but occasionally you will get a knowing smile from a fellow sheep hunter who has been there and understands. Just last summer I realized how much stairclimbing can affect leg strength. After changing my single-level home into a two-story home, my legs showed a noticeable improvement. And this was from climbing only one flight of stairs several times a day.

The author with his loaded pack after a successful solo hunt. Packin' will build strength needed to carry heavy loads.

By far the best way to get into "sheep shape" is by packin'. That is just what it sounds like. Get your pack and some sandbags (or any weight) and you are set. I've trained in this manner for about 20 years. Each spring I start out with a light pack and work up to a heavy one by fall. I like to get out early in the morning and walk for about an hour. I wear my hunting boots to accustom myself to their feel and work out any trouble spots. I usually go packin' three to four days a week during the summer months. In three or four months, I am usually ready. Lately I have discovered that I only need two to three weeks to work up to a heavy (120 lb.) load. Coincidentally, I have noticed some aches and pains associated with overuse of some of my joints, so I cut back on my training. I guess moderation is the key—particularly as we get older.

Hunters should also train their upper bodies for strength to get into their best "sheep shape". Arm strength is handy when the only way up a steep face is hand over hand through the alders. I remember the first bowhunt I went on when I ran into a vertical rock face. Luckily, I was able to use the overhanging alders like ropes and pull myself up and over the bare rock. There are a variety of situations like this when arm and hand strength can come in handy during a sheep hunt. The weight room can also be useful for upper body training, although it is not as important as with the lower body. Calisthenics and home gyms/apparatus can also provide all the upper body training you need. Good exercises for the upper body are plentiful whether you train with weights or just do calisthenics. Regardless of the exercises you choose, remember to select with your situation in mind.

Flexibility and Agility

Along with strength training you should include stretching exercises. Weight lifting has a tendency to tighten muscles, but even without weight training, our muscles shorten with use. You may not improve your flexibility by stretching, but you need to stretch just to maintain the range of motion you do have.

The author with his second bow-killed ram. Balance and agility are necessary to walk among the rocks and along the spines common to sheep habitat—especially with a heavy load. 38 1/8 x 14. Score--158 6/8 P&Y.

CONDITIONING

There have been many times during past sheep hunts when I've had to lie motionless in cramped positions while waiting for a sheep to move out of sight. I've also traveled many miles over the years in a hunched-over position to stay out of sight of a mountaintop ram. Having loose, pliable muscles is a valuable asset for sheep hunters trying to stay out of sight.

Stretching also helps prepare you for exercise. Whether you are going on one of your training runs or waking up in your tent on a frosty morning with stiff, sore muscles, stretching can help. Warming up by stretching allows you to work out harder and it can prepare you for a difficult stalk. Flexibility also helps prevent injuries. There is nothing worse than pulling a muscle or straining a ligament during a sheep hunt and having to turn back. Get into a habit of stretching every day to maintain your flexibility. Jack O'Connor hunted sheep into his seventies and, although I don't know this for a fact, I would bet he had a habit of stretching. The old adage "use it or lose it" is very appropriate when talking about flexibility and range of motion. If you always warm up before any strenuous exercise, it will save you numerous minor injuries over the years and extend the length of your sheep hunting career.

Flexibility also helps improve agility. Agility is obviously important to a sheep hunter, but it is difficult to work on specifically. Lunges are probably the best exercise to work on agility while improving leg strength. Agility will also improve as a result of all the other training I have suggested. Having excess strength will certainly help your balance as you walk down a narrow spine with a loaded pack. Stepping across loose, odd-shaped boulders so often found in sheep country also requires good agility—particularly with a loaded pack. The best agility training I can suggest is to actually get experience with a loaded pack on uneven ground. Your center of gravity changes as you carry heavier loads so experiment until you feel comfortable and stable. If you incorporate agility training into your packin' trips, you can improve agility as well as fitness.

Stamina

Stamina is the last parameter of conditioning you should work on. Stamina is the sum total of all the conditioning you have done. After you have been on an exercise program for several weeks and are in fairly good condition, work on stamina. For me this means having a longer and tougher workout. Maybe I will combine an aerobic workout in the morning with a weight training workout in the evening. Or, I will go on a three-hour packin' trip with some hills on the way. Pushing yourself beyond your normal workout will help build stamina.

Stamina is needed to continually travel up and down steep, rugged mountains where sheep live.

Perhaps the best way to build stamina is by taking scouting trips or training outings. Actually going into the mountains where your hunt will be is a great way to increase stamina as well as look over the country. Another bonus is that you can get acclimated to the thin air. Although Dall sheep are not usually found at extremely high elevations, the air at 4,000 feet is still noticeably thinner than at sea level. If you have been working hard on c/r. fitness the thin air won't affect you as much, but it's still noticeable. If you can't get to your hunt area in the off-season and there are no other steep mountains nearby, then just take a couple days and go hiking with a heavy pack. Find steep trails or at least long hills to climb. This will prepare you for the rigors of the hunt. The harder you push yourself in training, the more likely you are to be successful on the hunt.

You Are What You Eat

Diet accounts for 80% of the results of physical training. That is a statement I have read many times and didn't fully understand until I changed my own diet. Once I started eating a very good diet my conditioning program was much more effective. All the information we read and hear about eating whole fruits, vegetables, grains, and cutting down on fats is surprisingly good advice. If you have doubts about how significant diet can be, I suggest you try a **strict** low-fat, high-protein diet with lots of fruits, vegetables, and complex carbohydrates. Common sources of complex carbohydrates are grains, potatoes, and pasta. Sugars are simple carbohydrates and should be eaten in moderation. In just a few weeks you will be amazed at the effect a proper diet has on you.

If you are going to spend significant amounts of time on a physical conditioning program, eat sensibly to get the greatest benefit. And if you need to shed a few pounds, this is also the answer. But you shouldn't look at this as a "diet." Eating well should be a habit. Eating in this manner doesn't mean going hungry. I have a hard time not losing weight if I am eating right—even though I eat whenever I get hungry. I have to force myself to eat more just to maintain my weight. On the other hand, when I am not eating properly, the pounds always seem to gather on me. It is not the quantity of food we eat that puts on the extra pounds. It is those sugary and fatty foods that have so many concentrated calories that do it.

Actually, I have found it is better to carry a couple extra pounds of bodyweight when I go sheep hunting because it is such a physical hunt. In this way, I am carrying extra energy that will help me through demanding hunting days. It seems easier to carry this available energy on my body rather than in my pack. These extra pounds are typically gone before the end of the hunt. But I am talking about two or three pounds—not twenty or thirty. Decide how much you should weigh going into a sheep hunt and how important it is to get to that ideal weight. This is another decision to make about your situation.

While you are getting into "sheep shape" is a perfect time to test gear. Nothing is worse than having new boots and getting blisters after five miles of a 35 mile trip (as I have done—luckily it was just a scouting trip). Current footwear is supposedly broken in from the factory, but I still like to get new leather boots soaking wet and then walk them dry. And the new pack designs are great—**if** you know how to use them. Training walks will provide the opportunity to learn how to adjust them so you can be ready on opening day. Don't hinder your hunt because you don't know how to "operate" new equipment.

When testing new boots and packs, try carrying your gun as well (if you pick a suitable route for toting a gun). Toting a heavy gun while packin' heavy loads can be tiring. Training walks help determine whether or not to carry your chosen weapon over the distances required while sheep hunting. I once guided a 140-pound hunter who had absolutely the heaviest gun I have ever felt. I still don't know how he was able to carry that piece of artillery. He will think twice before choosing that gun for his next sheep hunt. Testing equipment beforehand can make a world of difference on a sheep hunt.

There is one more aspect of conditioning I would like to address--mental toughness. Stamina is a result of physical conditioning--mental toughness is a result of **all** the preparations you go through before a hunt. If you take the time to design a suitable physical conditioning program and stick to it, you will have confidence in your physical abilities. That confidence will improve your mental toughness and take you one step closer to a successful sheep hunt.

As I stated at the beginning of this chapter, these are only suggestions about physical conditioning. As your own personal trainer, first analyze your strengths and weaknesses. Then design a conditioning program based on your needs. Obviously, I am really "into" sheep hunting. I try to stay in good shape all year. I go to great lengths to tailor my physical conditioning program to improve my sheep hunting success. Of course, it is a good idea to exercise regularly just to maintain general health. Sheep hunting just gives me a good excuse to exercise year-round. You can pick and choose from the information I have given to design a program to get into sheep shape, or come up with other ways. My suggestions are not the only ways to improve physical preparedness for sheep hunting. These are just the methods I have come up with during my career as a sheep hunter. One thing is certain, the conditioning program you use will greatly affect your ability to hunt the white sheep.

I want to relate one last incident which stresses the need to get into top physical shape for sheep hunting. My neighbor took his

CONDITIONING

The result of proper conditioning and a lot of hard work. Mary Anderson with her ram. 33 x 12.

wife, Brenda, on her first sheep hunt a few years ago. Her comparison concerning the physical hardships of sheep hunting was "It's a toss-up between child-birth and sheep hunting." Since that time, however, she has had **two more** children, but has **not** been sheep hunting again!

Chapter 3

FOOTWEAR AND CLOTHING

Selecting gear for a sheep hunt is a monumental task. There are more innovations in gear than I have time to keep up with. I continually scan the catalogs and visit the sporting goods stores to stay informed about new products. Still, I know I'm missing out on some great new products that could improve my sheep hunts. As I try to stay informed of the latest gear, there are several qualities I look for.

Foremost in my mind is **safety**. Sheep hunting can often turn into a survival situation. Always keep that in mind when selecting gear. Sheep hunters are often many miles and days away from help. We must be self-reliant. Weather in the mountains is unpredictable at best and can be downright nasty for long periods. Torrential rains and strong winds are common. I have never been unfortunate enough to spend an entire ten-day hunt huddled in a small tent battered by relentless winds and unceasing rain. However, I have heard those horror stories often enough to prepare for it. The tent and clothing you choose will determine how comfortable and safe you'll be when the elements are against you. Keep that in mind when selecting gear.

The second priority when choosing gear is weight. Although I feel **safety** is the number one consideration, many hunters list weight as the top priority. Every mile traveled on a sheep hunt reminds you of that extra shirt or heavy sleeping bag in your pack. The more

Considering the country Dall sheep inhabit, hunters must place safety first on their list—don't you agree?

weight you carry, the less energy left to hunt sheep. I am constantly looking for a lighter sleeping bag or a way to carry less gear.

Also consider the space an item will take in a limited packbag. Sheep hunting gear should be as compactible as possible. Bulky items that must be carried outside a pack can be a real problem. They catch rocks, brush, and rainwater. And bulky gear makes it difficult to pack out a sheep in one load.

Other factors to consider are water resistance and how quickly an item dries. Water is both a friend and an enemy on a backpack hunt. You always have to plan on getting enough drinking water—while also staying dry. If your gear is wet, you will be uncomfortable at the least. If you are also cold, your **life** may be threatened. I choose gear that is either waterproof or water-resistant and dries quickly.

Reliability is another important quality of my gear. I can't worry about an important piece of equipment failing halfway through a hunt. I have heard dreadful stories about packframes breaking just as they were loaded full of sheep meat for the return trip. I have also seen cheap raincoats ripped to shreds after five minutes in the brushy mazes commonly encountered on sheep hunts. I suggest you buy top-quality gear.

FOOTWEAR AND CLOTHING

The final concern about gear is **cost**. Sheep hunters are notorious for spending huge sums of money on their gear. This is understandable when you consider the expense of a sheep hunt and the often once-in-a-lifetime opportunity it represents for many hunters. Sometimes I add up the value of the gear in my loaded pack. It runs into the thousands of dollars. The only reason the total is not higher is because I spend countless hours looking for the least expensive way to satisfy all the requirements of my sheep gear.

The details of the hunt will also help determine what gear to take. First of all, is it a guided or self-guided hunt? On a guided hunt, the guide should provide all the common gear like tents, stoves, spotting scope, etc. Communicate carefully with the guide about the gear you should bring. There is no need to duplicate gear. What is the duration of the hunt? How far will you travel? I am willing to take more gear on a shorter hike versus a marathon hunt. If I am going on a quick trip, though, I will often do without some comforts to save weight. Are you hunting solo or will you have partners? Is each partner planning to take an animal? The answers to all these questions significantly affect the gear you take. Careful planning can cut many pounds off your load and make for a more enjoyable and successful hunt. Careful preparation before the hunt is time well spent.

Footwear

There are two viable options in footwear for a sheep hunt: "plastic" or leather. "Plastic" boots aren't really made of dime-store material. These new climbing boots for hunting are similar to ski boots which are made of synthetic polymers. They are simply referred to as plastic because of their appearance. Leather boots are the traditional choice, but I would choose plastic boots for Dall sheep hunting in Alaska if I only had one choice. However, I use both—depending on the situation.

Plastic boots used for sheep hunting are made by several boot manufacturers. The best models have an outer plastic shell with tough lug soles. Inside of the shell fits a comfortable, insulated liner with a stiff sole. The liners can be conveniently removed for drying or worn around camp so aching feet can rest. They can also be used alone for the last few hundred yards of a stalk to reduce noise. The liner is made of non-absorbent insulation so it never seems to get as wet as leather boots. Even during the coldest sheep weather, damp plastic boots keep my feet warmer than dry leather boots.

33

photo by Dave Widby

Plastic boots are stiffer than most leather boots. This helps when climbing sheer walls, but is also important on steep, grassy slopes.

Another benefit of plastic boots is the shells can be used alone to cross streams. After a quick wipe to remove excess water, replace the liner and you can continue with dry feet. Plastic boots also cause fewer blisters than leather boots. I have only had two small blisters in several years of use and other hunters tell me of similar experience with their plastic boots.

The most outstanding feature of plastic boots is their effectiveness on steep slopes—they are **safe** to wear. I often come to steep faces with only narrow ledges for footholds that I wouldn't cross with leather boots. Wearing plastic boots, I often cross these same slopes with confidence. The stiffness of these boots allows me to get secure footholds where the more flexible leather boots won't. Sometimes just a steep, wet, grassy slope can be dangerous with leather boots; whereas the plastic boots will dig in and provide sufficient grip for me to travel in safety.

Another feature of plastic boots is their toughness. The hard outer shell protects my feet from rock bruises I sometimes get while using leather boots. While traveling the sheep mountains, I often ride steep shale slides as much as a thousand feet downhill. It is a great way to cover ground with little effort—particularly with a loaded pack. If I am wearing leather boots—I have to be concerned about bruises, but with plastic boots—I don't worry. This type of abuse also shortens the life of leather boots considerably; plastic boots hold up much better.

Now for the drawbacks. Plastic boots hurt my knees. They are heavier than leather boots and stiffer. The repetitive lifting and extending of my knees with these boots has given me an overuse injury of the knees. I still get sore knees in leather boots, but the heavier plastic boots are worse. It is also harder on my knees to walk long distances on relatively flat ground with stiff plastic boots. Leaving the top hooks unlaced gives me more freedom of motion so my knees aren't jerked as much with every step, but they are still harder on my knees than more flexible leather boots. Climbing hills with heavier boots is also more tiring. Plastic boots can also be clumsy and noisy, but both of these shortcomings can be partially overcome with experience.

I still use leather boots at times for sheep hunting. Before plastic boots came out, leather was the best option. They still are a justifiable choice. They are lighter and more waterproof now than they used to be. I have used the models lined with non-absorbent insulation and a breathable membrane for the past several years. They are great! The waterproof-yet-breathable membranes are very effective at resisting moisture. They are not perfect, but they are much better than only leather. Once the boots do get wet, they don't hold as much moisture, so they dry faster. The upper portion of mine are half Cordura nylon which makes them even lighter than before; yet they are tough enough for sheep hunting. I still would recommend waterproofing the leather as instructed by the manufacturer. I warm mine up and really work in the recommended dressing. This preserves the leather and improves their water resistance.

I have always used lugged soles for sheep hunting and I don't know of any other good choice. During a sheep hunt you will need to get good footholds on wet, slippery rocks. Lug soles are the best I have found for this. There are some new types of soles on the market, but I have not tried them. If you want to try them, do the proper research and testing on your own before trusting them on a sheep hunt.

The particular sheep hunt I am going on determines whether I wear plastic or leather. If there are long distances to travel over relatively flat and brushy terrain, I would rather have my leathers. If the hunt is going to be short and rough, I would prefer the plastics. If it is a late season hunt, the warmer plastics are more comfortable. If I am going with someone who is not used to climbing the sheep mountains, I can probably out-climb them with my leathers; so I can spare my knees the additional abuse the plastics would inflict.

The last concern about boots is cost. The plastics are generally 20-30% more expensive than **good** leather boots. Since footwear is so important, I generally disregard this difference.

Those are the characteristics to consider when selecting boots. If you are going on the hunt of a lifetime, wear plastic boots. You have a better chance at success with them. They are more likely to get you there and back in safety and comfort. Whether you decide on leather or plastic boots, I would suggest talking with several experienced hunters and knowledgeable sporting goods dealers before you purchase boots. The first-hand experience of hunters and the knowledge of dealers will help you make the right choice. Also, get a good pair of insoles for your boots. There are some great varieties available made of dense, shock-reducing materials that won't absorb moisture. These can really help prevent foot problems. They are definitely worth their minimal weight.

The other possibilities for footwear arise in a base camp situation. If you have mechanized transportation to a base camp and can afford the extra weight, bring a pair of camp shoes. It is a real luxury to relax at night in something other than your boots. I sometimes carry a pair of water-sport shoes with rubber soles and nylon uppers. They will keep your socks dry around camp and can also be used to wade streams. Bringing along such a luxury is the exception rather than the rule for me. My sheep hunts are almost always backpacking affairs.

Another option I have been toying with is using tennis shoes. On those hunts where the first part of the trip is going to be on a developed trail (a rare trip for me), tennis shoes would be faster and certainly less tiring. I have actually completed an entire sheep hunt wearing only a pair of sneakers. After losing one of my plastic boots during a river crossing, just after the plane departed, I had no choice. I was amazed at how fast the climb up a 3,000-foot hill could be without heavy boots to lift at every step. I think it would be worth the extra weight in some instances to carry the sneakers for the time and effort it would save in the long run. The sneakers could be stashed and picked up on the return trip. The characteristics of a particular hunt would determine if this was feasible.

Socks

Sheep hunters depend on their feet and it pays to take care of them. One thing I never scrimp on is socks. I always try to have several dry pairs. I take only synthetic socks. My favorite are high-bulk Orlon socks with double layers under the balls of the feet. Thor-Lo is the only manufacturer of these socks that I am aware of. They are expensive, but well worth the price. They are **great socks**.

Orlon is one of the acrylics that do not absorb much moisture, dry out quickly, and retain their loft. I can rinse Orlon socks in a creek, wring them out, and have nearly-dry socks immediately. Even in a damp tent in cool mountain temperatures, they seem to dry thoroughly overnight. I am completely sold on these socks.

I also wear some other synthetic socks. Polypropylene and Capilene are two more synthetic varieties of socks I recommend. Actually, both of these seem to be less abrasive next to the skin than Orlon but only marginally. I always wear two pairs of socks. An inner sock of either of these materials and Orlon outers works well.

The other options for socks are not satisfactory to me. Wool socks are the traditional favorite and they are warm. However, wool absorbs a lot more moisture and stays wet a lot longer than synthetics. It is also not as durable. You can get wool/synthetic blends that are much better than wool alone because they incorporate some of the best qualities of each material. These socks are okay, but I personally prefer the 100% Orlon. Cotton socks are worthless as far as I'm concerned, unless they are for a luxurious base camp—which I've never seen while sheep hunting. Cotton absorbs more moisture than even wool and seldom dries on a hunt. It also mats terribly. I leave my cotton socks at home when I go sheep hunting.

I always carry a pair of Gore-Tex socks. Gore-Tex is the same waterproof and breathable material that lines my leather boots. They are great! Although they are not completely waterproof, they do hold out water for a long time. Even if my boots are completely soaked and perhaps even frozen after a particularly cold night, I can still bring myself to don them with a pair of these "wonder" socks on. With dry socks inside a pair of these breathable socks, I can go for hours with soaked boots before the moisture begins to creep in. By then, my boots and feet will be warm and I hardly notice it. Before I discovered these socks, I suffered. I vividly remember having to walk in cold, **frozen** boots for hours until they thawed. I have even used plastic bags or stuff sacks for a liner between my feet and icy boots, but these new "wonder" socks do the job much better. My suggestion is—don't go hunting without a pair.

While I am on the subject of feet, I have some more suggestions. Always take Moleskin or a similar product. Earlier, when I mentioned that 35-mile scouting trip I took with blistered feet, I didn't relate how this adhesive material saved me. I ended up walking 30 miles with open blisters. The only way I was able to do this was with Moleskin over the hot spots. My feet basically didn't get any worse once I applied this adhesive bandage over the blisters. It is another invaluable item I never go without.

On the other hand, I never take gaiters. These are coverings that prevent snow from slipping down boot tops. Some sheep hunters use them to prevent rocks from working into their boots. I find their extra weight is not worth this minor utility. Plus, the nylon varieties are noisy. They have little value on a sheep hunt.

Finally, I want to discuss preventive medicine for your feet. All sheep hunters get blisters from time to time, even with the best boots and socks. It is part of the "experience." However, I try to minimize this aspect of sheep hunting. Over the years, I've tried everything to toughen up my feet. Nothing worked for me. I've also tried cooling them in icy streams to prevent hot spots and blisters from developing. Although I do soak my feet in icy streams from time to time because it feels good, it doesn't prevent blisters very well. The only way I have found to consistently reduce swelling of my feet and prevent blisters is by elevating them.

After I am through hiking for the day, I recline somewhere dry and raise my feet about 10 inches above the rest of my body for 15 minutes. After hours of abuse under a heavy load, feet swell. Without a chance to drain properly, either by resting for days or by elevating, feet are prone to soreness and blistering. I remember the aftermath of my early sheep hunts. I hobbled around on sore feet for a week or more before I could walk in a normal fashion. Now, I religiously raise my feet at night (daytime would probably work, but nighttime is more convenient) for 15-30 minutes and I seldom have sore feet. I count this as one of the most important lessons I've learned in my education as a sheep hunter.

Preventive medicine for feet. The author has found elevating his feet is the best therapy to minimize soreness.

Clothing

What I've learned about socks extends to clothing—I wear synthetics. I used to wear a lot of wool clothing when hunting. Wool has several desirable qualities and I still wear wool clothing for some outdoor activities, but seldom on a sheep hunt. The best choice in clothing for safety and comfort is synthetics.

A high priority for a sheep hunter is staying warm. If you are warm, you can avoid hypothermia and probably be comfortable, too. Hypothermia occurs when the body is unable to produce enough heat to prevent the core temperature from dropping. If this continues, the body dies. Moisture is a determining factor when losing body heat. To stay warm, it helps immeasurably to stay dry. That is the main reason I choose synthetics for my clothing. Synthetic fabrics inherently can't absorb much moisture. Thus, they dry quickly—sounds like my sock commercial, doesn't it? Synthetics also wick moisture away from your skin so you are drier and more comfortable. Also, because moisture is transported toward the outer layers of your clothing by synthetics, it dissipates more quickly than if it stayed next to your skin.

The other options for clothing are, again, wool and cotton. Cotton clothing is worthless on a sheep hunt. Once cotton gets wet (and it will get wet), you will be lucky to find enough time and sun to dry it on a week-long hunt. Don't endanger yourself—leave cotton clothing at home.

Wool hunting clothes are very popular. I own a lot of wool clothing and I used to wear it exclusively for my outer clothing on sheep hunts. However, the new fabrics are so much better I save my wool clothes for flatland hunting. Wool does repel water fairly well, but not any better than fleece does. Once wool gets wet (and, as I said with cotton, it will get wet), you will have difficulty drying it during a sheep hunt. I remember all too well the many times I have been cold and uncomfortable for the duration of a sheep hunt once my wool pants got wet.

Staying warm is vital to safety and comfort.

Also, wool is extremely heavy when compared to synthetics.

FOOTWEAR AND CLOTHING

Climbing the steep mountains that sheep inhabit is hard enough without having heavy wool pants holding you back. Wool pants don't give as you climb. They feel like lead weights to me now that I'm used to fleece pants. Fleece doesn't restrict my movements at all. Some wool garments do a fair job of stopping wind, while others don't. Wool is also bulky for it's warmth and, if you want good quality wool clothing, it is costly. I know there are a lot of experienced outdoors people who disagree with me about wool. I am not saying wool is worthless. Wool will always have a place in my hunting wardrobe. However, for sheep hunting, nothing currently beats the synthetics! Read on.

The synthetics I am referring to include polypropylene, Capilene, Orlon (remember my socks?) Thermax, ThermaStat, Coolmax, Aquator, Polartec, Worsterlon, and numerous others. There are many I haven't tried since they are coming on the market faster than I wear out the old clothing. These synthetics were developed for several characteristics. They are water-resistant or waterproof. Some new synthetics provide as much insulation per ounce as natural fabrics. Others are designed to be quiet, lightweight, comfortable, durable, and inexpensive. Of course, these characteristics are found in varying combinations, depending on the material you choose.

I have a pair of fleece pants that I have worn for four **hard** seasons. I am just now wearing them out. I have crawled countless miles on the knees and they only have one little tear in them—and it hasn't spread at all! They only cost about $30 and are extremely lightweight. They are also camouflaged and have zippers that go halfway up the legs. When the legs get wet, I simply unzip them and wring them out. They are dry almost instantly. They shed water like fur so I don't need raingear unless I'm in a downpour. They are the best pants to wear sheep hunting. The only valid criticism I have heard is they are not wind-proof. True. However, I can always put my rain chaps over them to stop the wind. If I was wearing heavier pants (like wool), not only would I get tired faster, I would also get a lot sweatier on the many climbs made during a sheep hunt. This would defeat the goal of staying warm, since wet clothing has less insulating value. If I need to match the warmth of heavy wool pants, I can take two pair of fleece pants and wear both if it gets really cold and windy. I would **still** be carrying less weight and have an extra pair of pants to boot. That brings up another axiom of sheep hunting—always dress in layers.

Several layers of clothing with warm air trapped in between them are warmer than the same total weight of fabric in one thicker layer—and more practical. Simply shed layers as needed to keep yourself cool—yes, I said **cool**. When you are traveling in sheep country, always try to stay cool—not warm. Once you have stopped for a while, then you can put on extra clothes to stay warm. The reason for this is to stay dry. If you are warm while you are traveling, you are often perspiring. If you can stay just a little uncomfortably cool, then you will stay much drier. It is always comforting to know you have dry clothes. Even though the synthetics dry quickly, they are still warmer and more comfortable when they are completely dry. Layering clothing is second nature to anyone who backpacks. It is necessary to stay warm (safe) and comfortable.

I wear a wide, full-grain leather belt to hang items on. I also wear suspenders. Suspenders give me a lot more freedom to move by taking most of the load off my belt. But don't get cotton suspenders! Find a synthetic pair. Nothing is worse than wet, cold suspenders against your back and they just don't seem to dry out. I have found good synthetic suspenders with black clips (which eliminate glare) at commercial fishing stores. The way to check if they are made of cotton or a synthetic material is with a lighter. If they burn—they are cotton, if they melt—they are synthetic.

My shirt is made of Worsterlon. I know that sounds like I only have one. Well, I only have **one** that I take sheep hunting and it is made of Worsterlon. Worsterlon is another "wonder" fabric and it is great for shirts. I have a pair of pants made of Worsterlon which aren't any good. They bind at the knee and resist me as I try to move. However, the shirt stops wind, sheds water, is tough as nails, lightweight, comfortable, and it costs less than a good wool shirt. I have found none better.

My coat is also made of fleece like my pants. It is light and has all the same qualities. I often take a vest made of the same material. It is homemade and only has Velcro closures. It weighs only a few ounces and provides another layer when needed. I also have a fleece shirt, but it doesn't perform as well as the Worsterlon shirt so it is left behind on sheep hunts.

I also take a light down coat or a Thinsulate coat for colder conditions. Down is extremely lightweight, but hard to keep dry. A Thinsulate coat is the better choice because of it's moisture resistance. Either way, another light layer is handy to have on cold, late hunts.

Sheep habitat is unforgiving. A sheep hunter must have the necessary footwear and clothing to survive.

I also recommend long underwear. A synthetic pair weighs only ounces and is very compactible. My top is made out of Thermax. It is very light-weight and dries very quickly. I can go to sleep while it is still damp with no discomfort. I often wear just this light undershirt when traveling. By doing this, I perspire as little as possible and don't get my shirt wet; yet I still have something between my skin and the pack to prevent blisters. My long underwear bottoms are polypropylene. The bottoms made of Thermax drag too much. The polypropylene doesn't restrict me at all. I have even gone to Thermax shorts to eliminate the last bit of cotton in my clothing. These are my preferences, but they may not suit you. Try out several synthetic fabrics and learn which ones perform well for you.

The last items on my clothing list are a hat, gloves, and a handkerchief. You need a good hat and gloves to survive comfortably in sheep country. We lose up to 50% of our body heat through our head and neck, so I suggest a full face mask. It is not only warmer than just a hat, but it is useful for peeking over ridges to look for sheep. A hunter's shiny face and neck have spooked more sheep than imaginable. By donning a face mask before looking over ridges, hunters significantly reduce the chance of being spotted. Wool hats are good and the synthetics are better for the same reasons as before. Gloves should be either wool or synthetic. I actually prefer wool because, when they get wet (and they always seem to get wet at least once), they keep me warmer than synthetic gloves. For this reason, I don't suggest lined gloves—they take too long to dry out and provide little warmth when wet. Wool or polypropylene can be wrung out as often as necessary and still keep you warm.

One item I don't like to be without in the mountains is my **cotton** handkerchief. The same quality that makes cotton unsuitable for sheep-hunting clothing makes it ideal for a handkerchief—it absorbs moisture. I use it for a scarf to keep my neck warm, a sponge to dry anything, and a washrag. On hot, sunny days early in the season a wet bandanna tied across my forehead keeps me cool. It weighs very little and, because it is so thin, I can dry it quickly. It is indispensable on a backpacking trip.

Raingear

Raingear should be lightweight, compactible, and effective. I say effective because no raingear is going to keep you completely dry on a sheep hunt. Rubberized raingear is the only type that will stop all the water from coming through. However, it is too heavy to carry on a sheep hunt. Additionally, you will get hotter and perspire more with rubberized raingear. Thus, you will get wet from the inside. The viable options are a high-quality coated nylon suit or one of the new waterproof-yet-breathable suits. Cheap vinyl or plastic raingear are poor choices; they are lightweight, but they disintegrate at the first sign of hard use.

Raingear made of one of the breathable fabrics can be fairly lightweight, but you need to shop around. I think the claim about breathability is justified to some extent, depending on which fabric you choose. As with most products, there is a wide range of quality among the many breathable fabrics available. However, even with the best of these fabrics, when hiking you will produce moisture faster than it can pass through the microscopic pores of the fabric. So, you will still get damp inside. Also, even the best suits of this type are not completely waterproof. In a heavy rain you will still get wet from the outside, albeit slowly. Suits made of these fabrics are also fairly compact. Their big drawback is price. I have never owned a suit like this myself because they are so expensive. If you are in the market for one of these suits, I would suggest using the same method as I described for choosing footwear. Ask hunters with Alaskan experience and reputable dealers to gather information before you buy.

My choice in raingear is a suit of polyurethane-coated taffeta nylon. Taffeta is a very lightweight nylon and, when it is coated with polyurethane, it is absolutely tough as nails. It is about 95% waterproof and dries quickly. My rainsuit will keep me fairly dry, unless I'm in a driving rain, and I don't get too hot if I have to wear it while hiking. The cost is low and it is very compactible. The one drawback is availability. I have seldom found rainsuits of this fabric that are also camouflaged and designed properly. Suits in blues or reds are available, but those colors aren't good for sheep hunting. My solution has been to make my own. Unfortunately, the material is also hard to find in camouflage. When I do find the right material, I buy a large supply. These suits wear out in 3-4 years of hard use, but replacing them is inexpensive—if you have the material. By making my own rainsuits, I also get the exact design I like.

I am very particular about the design of my raingear. I prefer a loose raincoat with **large** pockets. I like a tight-fitting hood that can be closed to a small opening and protect my face. I also like Velcro around the wrists to keep rain out. I want the coat to come down to my knees so I can sit on wet ground and stay dry. It has to be long because I don't wear rainpants—I wear chaps. I make rain chaps out of the same material as the coat, but they are available in camouflage through several catalogs. They weigh almost nothing and can be put into a small pocket. Rainpants always make me too hot and sweaty. I find that I get much better ventilation by wearing chaps and I don't perspire as much. The result is I stay drier, which is my main goal.

Selecting footwear and clothing for a sheep hunt is a difficult task. There are many options. Most are expensive. Well-informed sheep hunters can make wise decisions while keeping cost down. The next chapter deals with an equally important and challenging topic—**backpacking gear.**

Chapter 4

BACKPACKING GEAR

In addition to suitable boots and clothing, a sheep hunter needs appropriate backpacking gear. Remember to look for gear that will keep you warm and dry. Ideally, sheep hunting gear should also be durable, lightweight, and compact. The trick is to satisfy all these requirements while minimizing costs.

Backpacks

Before choosing a backpack for a sheep hunt, the questions about time, distance, number of partners, and guided versus self-guided hunt need to be answered. These answers will help determine the load requirements of a pack. The size of the pack bag, the weight the pack can carry, and the comfort of the loaded pack are the most significant characteristics to consider.

The two options for a pack are external frame and internal frame. In general, you can carry more weight with an external frame pack. If you are planning a solo hunt or a hunt where each partner is expecting to take a sheep, take an external frame pack. The newer internal frame packs are far superior to older models and backpackers with strong upper bodies can carry 80 to 90 lbs. comfortably with them. However, a pack loaded with a sheep and hunting gear will weigh in excess of 100 lbs. If you are going on a guided hunt or a hunt where two people will pack only one sheep, then an internal frame pack may work. Whichever type of pack you choose, I suggest getting as large a pack bag as possible. The extra room is always useful yet adds hardly any weight.

With an external frame pack, "sheep-shape" hunters can carry their gear plus the meat, cape, and horns.

External frame packs allow you to carry more weight because they are stiffer. This stiffness also prevents the pack from flexing as you bend. Consequently, it is more tiring to climb steep hills and maneuver through brush. An internal frame pack is much more comfortable to me, if I am only carrying 40 to 50 lbs., because it allows me to move through alder mazes and around steep cliffs with much less effort. I can also maintain my balance much better using an internal frame pack with a light load. It flexes with me instead of resisting as an external frame will do. However, I sweat more with an internal frame pack because it is tight against my back. The external frame pack allows for a little more ventilation between my body and the pack.

Both pack types can be found with large capacity bags and plenty of pockets. Make sure the pack is durable enough to handle the rigors of a sheep hunt. Any mountaineering-quality pack should do. The pack bag should also be coated to provide some water resistance. I often carry an additional waterproof pack cover, but it is not necessary if you have a good pack bag. As a pack gets more

BACKPACKING GEAR

use it will leak more—which is when the pack cover is needed.

Look carefully at the suspension system when choosing a pack. Look for well-padded shoulder straps, a wide hip belt, and a chest strap. A good suspension system allows you to carry 20 to 30 lbs. more than a poor one. The straps should be solidly attached to the pack. Test the grommets on the bag to convince yourself that they will not fail halfway through a hunt. External frame packs usually employ aluminum pins and retainer rings to attach the bag to the pack. I suggest carrying a few extras as they sometimes break or come off.

Before you buy a pack, try it on. Packs come in different sizes for different size people and they are adjustable. Find a pack that is comfortable for **you**. Then adjust it carefully to fit your body.

Tents

Your tent can literally save your life. It is a most important survival item. Rain and wind are the two main contributing factors in exposure. You must have a good tent (or shelter) to protect yourself from them. Choose a tent that sheds both well. Even if you are not faced with a survival situation (though most sheep hunters frequently are—even if they don't realize it), a good tent will make your trip more comfortable. And, a well-rested sheep hunter is more likely to be successful.

A good mountain tent needs to be breathable **and** waterproof. Most backpacking tents are currently designed with two layers—a breathable inner tent and a waterproof outer fly. The upper portion of the inner tent should not be waterproof. This will allow some of the two quarts of water you and your partner exhale every night to pass out of your sleeping space. The floor of the inner tent and the entire fly must be completely waterproof.

Tents are also available which have only one layer made of the waterproof-yet-breathable fabrics I've mentioned before. They are slightly lighter than traditional backpacking tents with two layers. They are also more expensive and they sweat more inside. I don't think the weight savings is significant enough to justify the extra moisture and expense.

A good mountain tent is aerodynamically-designed to shed wind. This can be determined just by looking at the shape of a tent. Any vertical wall or high spot will catch wind. Thus, you need to look for a low-profile, streamlined tent. You won't have much headroom inside, but that is a luxury sheep hunters can do without. A two-man tent of this type should weigh about four pounds to qualify for sheep hunting. You will pay a good price for a suitable tent, but it is worth it.

One of the best ways to compare features among tents is by looking at rating sheets. Several catalogs contain them. These comparison sheets sometimes even rate durability in high winds as well as interior space, weight, height, packing size, and price. By using these comparison sheets, you can quickly look at the features of several tents. Once you find the right tent, shop around for the best price.

A windbreak like these rocks also provides cover from nearby sheep.

Test your tent before actually going on a hunt. You may learn some very useful information that will improve your hunt. For example, my tent has been used in high winds several times. It will only withstand about 35 mph winds before the poles start to bend. However, I have discovered it will withstand twice that velocity of wind if I lay one of the support hoops down. It may not be as comfortable with the tent constantly flapping in the wind nor quite as waterproof, but I have slept through a couple nights this way that otherwise would have been spent in some less-desirable fashion.

The location and the manner in which you pitch a tent are as important as it's design. The first thing to consider is wind. It is always a good idea to pitch your tent next to some kind of windbreak. This can be a convenient rock outcropping or bush, or it may just be a rise in the tundra that will lift most of the wind over your tent. Try to imagine what may happen if the wind comes up suddenly— a frequent occurrence in the mountains. Avoid any spot that looks like a wind funnel. I learned that lesson long ago in the Alaska Range. We placed our tent in a very attractive, level spot. Unfortunately, we were in the middle of a saddle connecting two large drainages. Luckily, when the wind picked up from zero to fifty mph in the span of ten minutes, we were still setting up the tent and not already inside. We flew down the side of that saddle with the tent acting as a sail between us. The outer edges of picturesque promontories are also good places to avoid for the same reason.

Most backpacking tents are designed with a gradual slope at one end. This end should be placed into the wind or quartering into it to reduce drag.

Pitching a tent on a gradual slope is often a good idea to avoid hidden water frequently found in level spots. It is often impossible to even find level spots big enough for a tent in sheep country anyway. Experience with your tent will determine how sloped the ground can be and still be usable.

It is a good idea to **always** anchor your tent with rocks. Not only will this frequently provide a better night's sleep, it will reduce the chances of returning from a day's hunt to find the wind has relocated your tent.

When close to your quarry, another possibility to bear in mind is getting pinned down in your tent. If possible, plan an escape route in the event that you awake and find rams nearby. This is not a remote possibility at all; most sheep hunters have had this happen. Although it makes for a good story, it can be frustrating. In most cases, I try not to pitch a tent within a mile of sheep to avoid this.

Sheep don't seem to be disturbed by colorful tents. I don't worry about the color of my tent for this reason. I even have appreciated the off-color of my tent when returning after dark. It is no fun hunting for your tent when you are wet, tired and every step is an effort. I also find that having a brightly colored tent will alert other hunters that someone is already in the area. Often, they will then move on to another valley.

Sleeping bags

Your sleeping bag is another item that you will depend on for comfort and safety. Having a dry sleeping bag can make the difference between success or failure (and life or death). It takes a very tough sheep hunter to continue hunting once their sleeping bag gets wet. If your sleeping bag is down-filled and it gets wet, you will have a hard time staying warm. The synthetic-filled bags are the solution to wet weather typical of sheep season.

There are dozens of good synthetic bags on the market. All the different synthetic fill materials are non-absorbent and keep their loft even when wet. So, even if your bag gets completely soaked, it can be effectively wrung out. Although you may be uncomfortable in the damp bag, you will probably survive.

Shop around to find the best bag for you. Like backpacks, sleeping bags come in different sizes to accommodate the various sizes of people, and proper fit does make a difference. There are also comparison guides for sleeping bags which are very helpful when selecting a bag. I try to find a bag rated to about +10° F. This seems to be enough for our sheep season here in Alaska. If it is colder, I can go downhill and build a fire—or walk out.

There are ways to stay warmer in an emergency anyway. Any insulating material can be used. I use all my clothing and gear inside the tent at night to be comfortable. I lay it under and over me depending on the situation. If dead vegetation is available (like grass or fireweed), it can be used as padding under you or over you for insulation.

Since I am very weight conscious, I don't carry a bag that is rated for colder temperatures because it will be heavier than necessary. My synthetic bags for sheep hunting weigh two to three pounds. Generally, the lighter ones are more expensive, so you should shop around to get the best deal. Synthetic bags are not quite as compressible as down bags, but the new ones are pretty close in both bulk and weight. And, the price for synthetics is usually less. The water resistance of synthetics and their insulating qualities when wet make them the best choice. Always carry your sleeping bag in a waterproof stuff sack. It is **well worth** the extra ounce to have a dry bag.

The author used all his clothing and gear to stay warm enough to continue this hunt—it paid off! This ram is currently the #1 Pope and Young Dall sheep. 42 4/8 x 14 2/8. Score—171 P&Y

You may want to carry a sleeping pad. There are many good ones on the market. I like the closed cell variety. I cut it so it only reaches from my hips to my shoulders. I stuff clothing in one of my stuff sacks to put under my head and I put my excess clothing under my legs and feet. I put my raingear down first as insurance against moisture and I even use my game bags under me. If time permits, I try to find dead vegetation to put under the tent and often have 6 inches of padding, even after sleeping on it for a few nights. I think a longer pad is just unnecessary weight. My pad weighs just 5 ounces.

Some of the inflatable types that, admittedly, are comfortable, weigh over two pounds!

Stoves

A good backpacking stove is light, small, stable, efficient, heats quickly, and functions well in cold weather. Good stoves may seem expensive at first, but they are sure appreciated at mealtime when ravenous sheep hunters want food. MSR stoves are one brand that offers several models which satisfy all of these requirements. Additionally, the MSR stoves use separate fuel containers. Therefore, various size fuel bottles can be carried depending on the duration of the hunt. The model I use weighs under 16 ounces and still costs less than $50. I have recently noticed a new stove on the market that is rated as highly as mine, plus it is cheaper **and** lighter. As I have said, with all the new gear being developed you should continually watch for better equipment. Look over the **current** catalog rating guides for stoves before choosing one.

Cookware

Most of a sheep hunter's food is either ready to eat or only requires the addition of boiling water. Consequently, only a few utensils and a small amount of cookware are needed on a sheep hunt. I carry one pot, a Large Lexan cup, and a Lexan spoon. I can cook and eat anything with these—along with my hunting knife. My pot holds my stove while I am traveling to conserve space. It is just big enough to boil water for two people at mealtime. I use a Lexan spoon and cup because they don't get too cold or too hot and are lighter than metal. If a base camp is used another larger pot may be added to this list; but that is the only change I would make.

Water containers

One of the primary concerns on a sheep hunt is finding enough drinkable water. Once above treeline, water can be hard to find. The top of a sheep ridge can be like a desert. It is no fun to descend a thousand feet or more just to replenish your water supply. Therefore, I always carry two water bottles. I don't always keep them full if I know I will be able to get more as I need it. But, if I am at all uncertain about finding a water supply, I fill both of them at every opportunity. I learned that lesson the hard way on my first sheep hunt. My brother, Randy, and I had just taken a sheep on a steep, rocky face. We spent three hard hours just getting to the sheep by a roundabout route. We foolishly passed up water sources on the descent so we had only a quart between us. After loading the sheep and starting the climb to camp, we ran out of water. Though we were hungry, I remember being so dehydrated I could not swallow food. We went up the mountain so slowly that darkness caught us on an exposed ridge. We spent a very uncomfortable night huddled among some rocks. Luckily, we survived, but we were prime candidates for hypothermia. We had no tent or sleeping bags and our dehydrated bodies could not function at full efficiency to warm us. After finding water the next morning, I made a vow never to go without adequate water again. The extra quart bottle weighs 4 ounces, but I have not been able to justify leaving it behind.

Drink your fill at every opportunity. Sheep hunters need enormous quantities of water to operate at 100% efficiency. Try to drink at least four quarts per day!

Your body requires enormous quantities of water while backpacking for sheep. You will be burning vast quantities of energy as well as perspiring profusely. Both drain the body of water. Plan ahead as much as possible to have plenty of water. If the human body is down just a quart of water, it will operate at less than 100% efficiency. If it is down two quarts, you will be operating at only about 80%. Drinking four quarts of water a day might be enough, but more would be better. Whenever you pass drinkable water, drink as much as possible. It seems like a nuisance at times, but it will pay off in the long run.

To locate water in the mountains, look for shiny spots on hillsides or listen for the sound of running water. I have found water by removing two feet of large rocks covering a trickle of water that I could just barely hear. I have also used a rain fly to catch water when the nearest source was two thousand feet below us. Even a light rain will produce amazing amounts of water in a short time. Noticing where water might be often helps determine my route in the mountains. It becomes second nature and it makes for a safer, more enjoyable hunt.

I need to say a few words about the potability of water. Even in Alaska **all untreated water must be suspect.** Basically, you can never be 100% sure any natural water source is safe to drink. Even a source coming right out of a mountainside can be contaminated by the animals living there. The only way to have 100% safe water is to treat it by filtration, boiling, or adding chemicals. The newer, compact, water filtration systems are fairly lightweight and reportedly 100% reliable. Traditional boiling or chemical purification will also work if done properly. There are many sheep hunters who just take the risk and don't use any purification methods. I can't recommend this as it is not safe. You need to select a method that satisfies your needs.

Climbing poles

I carry a three-section, aluminum, collapsible ski pole when I go climbing. The pole makes mountain traveling safer and makes me more efficient. It weighs in at 11 ounces and I would not go without it. Using the pole gives me one more point of contact with the mountain. When crossing a shale slide, I can plant the pole uphill and have **two** points of contact with the rock—one foot and the pole. Two points of contact make me much more stable as I look for my next foothold. Imagine balancing on one foot with a 100-pound-plus load as you search for a good solid spot for your next step among loose, irregular rocks of all sizes. You must quickly find a solid place for the next step—before losing your balance or your present foothold gives way. This is just as risky as it sounds. However, with a climbing pole, you can take your time and carefully pick the next step.

A climbing pole not only saves mishaps that may result in injuries, it also saves wear and tear on your knees and legs. Whether going up or down, the pole allows you to use your arms and take some of the weight off your legs.

The pole can also be used to cross streams. Rather than trying to find a suitable stick at every stream crossing, or just taking a risk and going without one, you will have the pole to rely on.

With one of these poles, hunters travel with less effort and more safety. The weight of the pole is more than compensated for in the energy saved and the security it bestows. **All sheep hunters should carry one.**

First Aid

Hunters should always carry a first aid kit. It does not have to be big—mine is only 4 ounces. I carry common items like Bandaids, aspirin, Moleskin, an anti-diarrhetic, and an antibiotic cream. Mine is probably a little skimpy for most people. Each hunter should make a kit to satisfy their needs. Make sure to carry ample prescription medicine in case your hunt is extended beyond what was planned. Keep in mind that all medical emergencies will have to be solved with your kit. Help may be many miles and days away if something happens. Plan accordingly.

Another item I don't go without is my elastic knee bandages. They are simple tubes of elastic that surround the knee and act as another supporting ligament. They are not so much for injuries (although they have come in handy more than once—and not just for me) as they are for tired legs. Whenever I am coming downhill with a load, I put them on and they save me a lot of effort. Instead of your leg taking the full shock of each step, the braces help stop the combined weight of you and your pack. They are definitely worth the 3 ounces they weigh.

Etc.

There are some other necessary and/or helpful items for a sheep hunt. I always carry two knives—a short sheath knife and a pocket knife with two small blades I use for caping. I feel the extra three ounces the pocketknife weighs is justified by the advantage in caping and the security of having an extra knife in case I lose one. I make lightweight game bags and carry one for meat and one for the cape. I carry a small half-ounce steel that is sufficient for sheep hunting. You shouldn't have to sharpen your knife at all on most sheep hunts and, if you do, a small steel will suffice. I have seen some huge steels in hunter's packs that I would not even carry on a flatland hunt. You also need one boning saw for your hunting party to remove the horns. Mine came with a leather sheath that I immediately replaced with a nylon copy to save weight.

    Hunters should also carry a compass, maps of the area, two lighters, waterproof matches in several pockets, and firestarters. I have a small plastic flashlight that conveniently indicates the amount of battery life remaining. I like plastic because it is more comfortable if I have to hold it in my mouth while using both hands. A small mirror is very useful to remove objects from eyes and for emergency signaling.

    I also carry a lightweight (7.5 ounce) 35mm camera. I always take two rolls of film—one roll of 100 ISO and one of 200 ISO. After putting in all the effort it takes to complete a successful sheep hunt, you will want good photos of the hunt. Although not always convenient, the effort spent to take numerous pictures is worthwhile. The 200 speed film really helps on overcast days or at dusk. It is also useful to have a timer on the camera so you can take your own picture (if you hunt solo like I usually do) or get all of your group in one photo.

    I always have at least one plastic bag for various uses. It's primary use is to hold the meat bag. This prevents blood from getting on my gear. Of course they can be used to keep almost everything in your pack dry when necessary. Plastic bags are handy to have, but remember, they are also more weight to carry.

    I also carry 20 ft. of light rope lashed to my pack and enough twine to tie bulky gear outside my pack once I fill the bag with sheep. Other handy items I carry are rubber bands, note paper with

BACKPACKING GEAR

a pencil in a plastic bag, and a little tape wrapped around my toothbrush. Personal toiletries are also included.

Photographs should be used to record all portions of a hunt. This view graphically portrays why safety is a prime concern. A snowstorm has made the return hike down this icy mountain a dangerous task.

Organization

Once your gear is assembled, you can start to reduce the pile. Even after years of experience there always seems to be too much stuff in my pile. I'm sure I still take unnecessary items or more of something than I really need. There are numerous ways to reduce the amount of weight you carry on a sheep hunt.

One strategy is to use items for multiple purposes. I use all my clothes and several items of gear for padding and warmth at night. A pack can even be used as one booty to warm the lower legs. A tent rain fly can be used to keep warm on a windswept hillside.

You can also alter items to reduce weight. I leave heavy cases (camera, saw, stove, etc.) behind and make lightweight replacements if necessary. Some sheep hunters (fanatics?) have been known to cut their toothbrush handle in half and drill holes in other items to reduce weight. I won't **admit** to such extremes, but the weight of every item you carry should be considered.

You should also learn how to pack gear so it is easier to carry. Put the heavy items midway up the pack and close to your back. Lighter gear can be placed in side pockets or high in the pack. Keep most of the weight low and in the center of the pack. By doing this, you can maintain balance with a lot less effort. This is not so crucial when carrying only 40 to 50 lbs., but it will be when packing 100 lbs. or more.

Camouflage

I use camouflaged gear. The soft items that aren't available in camouflage can be painted with fabric paint. This will at least break up any solid colors. If an item has any shine I paint it or cover it with tape. Look over your gear to see if anything will stand out in the mountains. Sheep can see long distances—I have seen them leave the country after seeing just part of me at over a mile. Camouflaging your gear reduces the risk that an unseen sheep will spot you. It is worth the effort. I have to repeat paint jobs on some gear frequently as the paint wears off with use, but it is easily done.

My last suggestion about gear is to have an "extra" bag. I take a waterproof, nylon bag containing a few items to leave at the airstrip. This can make waiting for the return plane a little more comfortable. If you are going to leave mechanical transportation at some point away from a road, it is convenient to have some extra gear there. I

BACKPACKING GEAR

fill mine with extra ammunition, durable fruits and vegetables (like apples and carrots), other dried food in case the plane is late, an extra tarp (to sit and tell stories under when it is raining), and other items that aren't too heavy. If you can afford the extra weight, I think it is a good idea.

A provisions bag left at an airstrip can make waiting for a plane much more comfortable. Here the author waits after taking his third Pope and Young ram. 40 1/8 x 12 6/8. Score—151 6/8 P&Y.

Chapter 5

WEAPONS AND OPTICS

I have read more about these two subjects than I can possibly remember. I am certainly not an authority on cartridges nor ballistics. Like most hunters, I do not have the time or inclination to learn everything there is to know about weapons. Digesting technical information about optics also gets tiring. However, I have spent enough time educating myself by reading and through extensive field experience to know what works for sheep hunting.

Firearms

The ideal rifle for sheep hunting is light-weight, flat-shooting, and accurate. Sheep live in open country, and sometimes the only possible shot is at long range. Shots at 300 yards and farther are not uncommon. A Dall ram weighs only about 200 lbs. and they are not particularly hard to kill. The most common cartridges are in the 30 caliber range—the 30-06 is a perennial favorite. The .270 meets all the requirements of a good sheep cartridge and can be made into a light-weight rifle. Most sheep hunters try to keep their rifle under 7 lbs. This is much easier with smaller calibers.

The only drawback to using such a small caliber is the presence of grizzly bears in sheep country. Hunters must weigh the advantage of a lighter rifle against the need to stop a bear if necessary. Personally, I have never had a bear problem while in sheep country, but I still use a magnum cartridge.

Dall sheep hunting is often a **wet** affair as I have stressed. Therefore, you need a rifle that will hold it's zero in sloppy conditions. Using a synthetic stock or glass-bedding the barrel should insure this. Synthetic stocks also reduce the weight of the rifle, though their appearance is offensive to some hunters. A stainless-steel or other corrosion-resistant barrel would also be easier to protect against moisture.

Even with all the recommendations for the "ideal" sheep rifle the one you use will reflect your priorities. Presently, I **hunt** with a bow. However, I also **guide** sheep hunters and need to carry a rifle at those times. It is most important for me to have a large caliber with sufficient knockdown when I am with a client. I use a .338 Winchester magnum. It is the only large-caliber rifle I own and I have used it for everything here in Alaska. The main reason I keep using it is for bear protection. However, my .338 is also flat-shooting and I am confident I can shoot it well. In my opinion, marksmanship is just as important as caliber selection.

Although some rifles can be shot accurately out to 500 yards, very few hunters should try such a long shot. I can't personally shoot accurately at that range even if I had a weapon capable of it. Most hunters cannot judge distance well enough to shoot accurately beyond 200 yards. Some should limit themselves to 100 yards or **less**.

Put a good sling on your rifle. There will be many times when both hands will be needed to climb up or down a steep face. There will also be circumstances when you should travel in a crouch. In either case, a sling will be convenient. I use a nylon model because it is both light-weight and doesn't absorb much water.

Due to the long-distance shots common to sheep hunting, most hunters use rifle scopes. I use a 4x. Many experienced sheep hunters use stronger scopes and even variables. There are times when being able to raise the magnification of a scope to 9x may help. However, variables don't always hold their zero when the power is adjusted and they are heavier and more prone to failure. Finding a model you can trust in the inclement weather is imperative. A fogged scope is useless. The mounts should be solid and tight. It is a good idea to put something like Loctite on the screws to minimize the effects rough treatment can have on your rifle's point of impact. One more recommendation—use good scope covers to keep water and debris off the lenses. As insurance, you can carry a small chamois cloth in a Ziploc. Use it to dry scope lenses at critical moments.

It is a good idea to periodically check the zero of your rifle while sheep hunting. Weapons get a lot of physical abuse on this type of hunt. It pays to take a few minutes occasionally to check their accuracy. This can be done by bore-sighting. Bore sighting is as simple as placing the rifle on a solid rest and then looking down the barrel and through the sights to ensure they point to the same spot. While this method will not indicate if the sights are only slightly off, it does make gross misalignments obvious. I can personally adjust my rifle's point of impact to within an inch or two of zero at 100 yards by doing this. More than once I have dropped my rifle among the rocks and had to check it's zero in this manner. Even though I have seldom found my rifle to be off after such abuse, it is comforting to know for sure it is still shooting where I expect it to.

It is hard to keep rifles free from rust on an extended sheep hunt. I usually place a small piece of tape over my barrel to keep moisture and debris out. I also puncture the tape because I am not convinced it will not cause excessive pressure. My cleaning kit consists a small, oil-soaked rag and a string. I clear the barrel by pulling the string through it after tying a piece of cloth on the end. There are ultralight, compact cleaning kits on the market which are more versatile then mine. Your cleaning kit will depend on how moisture-resistant your rifle is and how scrupulous you are about taking care of it. Regardless of how careful you are, expect to acquire some souvenir scratches on it from a sheep hunt.

More important then the selection of your rifle is the amount of practice you get with it. Most of us are limited by our skill and not by our weapons. Practice will educate your trigger finger and provide an opportunity to find a load that suits **your** gun. Every rifle is different. Even rifles of the same caliber and model don't shoot a given load with the same accuracy.

While looking for the best load, examine ballistic tables to compare trajectory and energy. Knowing what your weapon will do at different ranges will help decide where to hold and when not to shoot at all. The first time most shooters look at their bullet's long-range trajectory they are amazed at the amount of drop it has.

Practice should also include estimating ranges. I often guess ranges to telephone poles and animal-size objects as I go on conditioning walks or even walking city streets. This kind of practice can really pay off when you are in the field with only one chance to make a good shot. Knowing the trajectory of your bullet is useless unless you can accurately judge distances in the field.

There is one more item of concern when determining where to hold on a distant sheep. I know this has been written many times, but I am going to repeat it. When shooting uphill or downhill, the amount of drop your bullet exhibits is only affected by **horizontal** distance traveled. If a ram's position is uphill **or** downhill from you, it makes no difference which when determining point of aim. If a ram is directly below or directly above you at any distance, gravity will not cause a bullet to curve at all—aim as if he is only a few feet away. Practice will demonstrate this.

Archery equipment

You need to consider distance of shots, weather, and terrain when deciding on proper archery gear for a sheep hunt. A flat-shooting set-up is helpful because of the difficulty of getting close to sheep. An overdraw with light arrows improves your chances of making a longer shot, but they are also more difficult to shoot accurately under field conditions. Light arrows are also more susceptible to erratic flight in the windy conditions often found in the mountains.

The wet weather encountered on most sheep hunts is also a consideration. Wood limbs and arrows can be difficult to protect from the moisture. If using either, watch for delamination of limbs and warpage of arrows. Current archery tackle is generally made to withstand moisture, but any wood components should still be checked frequently. I suggest non-wood limbs (if possible) and aluminum arrows to minimize moisture problems. I remember the first bowhunt for sheep I went on and the inclement weather. After 5 days of rain I awoke one morning to several inches of snow and a bow that was covered with solid ice. The snow continued to fall as I hunted that snowy morning. I had to continually clear my shafts, nocks, and broadheads of ice and frozen snow. Despite the conditions, I was successful that day. I doubt wooden components would have performed as well in this situation.

Since sheep hunting is usually a backpacking affair, plan on carrying your bow and arrows over many miles of rough terrain. I use an arrow tube that holds 8 arrows. I made it from a fishing rod case, but commercial varieties are available. I have a sleeve sewn on my pack to carry the tube. Once I begin hunting, I leave it at camp.

George E. Mann moves across a rain-soaked shale slide. Rough terrain and inclement weather are hard on equipment.

I also prefer a short, light bow for carrying and hunting in the rocky terrain. Bows can be strapped to the back of a pack with a Bungee cord while traveling. This will leave your hands free. A short bow is easier to handle on a pack and easier to shoot on steep, rocky slopes. Some compound bows are designed so you don't even need a bow press to take the limbs off. Once you are done hunting it is very convenient to take the bow apart and put it inside your pack. One hazard I have encountered by strapping a bow on my pack is it can bump into objects behind me and throw me off balance. That is the last thing you want to happen on a steep cliff with a heavy load. The minor inconvenience of re-stringing and re-tuning a bow is worth the benefit of being able to pack it away on the return trip.

A light-weight bow is also much easier to pack than a heavy one with all kinds of attachments. Bowhunters don't usually consider weight when they pick their bow. However, it is an **important** consideration for backpacking hunters. The quiver and stabilizer can be removed to make a bow more compact for packing. A takedown recurve or longbow would be ideal for hunting sheep if you can shoot it well. I recently acquired a custom take-down for just that reason.

Check your bow, quiver, rest, sight, etc. for glare before the hunt. Don't risk spooking sheep at long distances by carelessness. Bowhunters should also have a bow sling. Stalking sheep usually involves climbing while staying low to the ground. It helps to have both hands free. I have crawled up to a half-mile while on a stalk. A sling makes this **much** easier.

Sheep hunting is tough on equipment. Always carry an extra string. Tape it to your quiver where it will always be readily available. Don't count on using any arrow or broadhead twice, even broadheads you can re-sharpen If you do find arrows after shooting them, they are usually broken, bent, or without points. Take as many arrows as you expect to shoot—only one shot per arrow.

Bowhunters need to practice a lot more than gun hunters. A bowhunter after sheep needs additional practice at steep angles and across open spaces. Practice judging distances over gullies and ravines just like an actual sheep hunt. It is much more difficult to accurately judge distance over a ravine than flat ground. Go into the mountains to do this. It's the only way to get comparable practice. You can check your estimations with a rangefinder. Practice steep angle shots to see how your arrow responds. Learn how to keep the arrow on the rest at all angles and how different angles affect arrow placement. This type of practice will demonstrate how an arrow's trajectory is affected when shooting uphill or downhill. Like rifle hunters, bowhunters should also aim with only the **horizontal** distance in mind. This is even more crucial to bowhunters since our point of impact is affected significantly by only a few yards of distance.

As mentioned earlier, sheep are not particularly hard to kill, so any good, **sharp** broadhead will work. I carry 9 Thunderhead 125's in a plastic, water-tight box. They have always performed well for me. Use whatever you shoot **accurately** and have **confidence** in.

Even recurve shooters have had success taking Dall sheep. Bart Schleyer is understandably pleased with his ram. 34 x 13.

Optics

It might seem like an easy task to spot white sheep on a dark hillside—and it usually is. But not always. When Dall sheep are in shadows they can be as difficult to spot as any dark animal. It can also be tough to determine if rams are legal and/or worth going after from long distances. Consequently, one of the most important pieces of equipment you carry on a sheep hunt is a spotting scope. Binoculars can be useful, but are often left behind because of weight considerations.

Binoculars are optional if you carry a quality spotting scope. A variable-power scope adjusted to it's lowest magnification can be hand-held to check for sheep and/or horns. However, it can be a nuisance to repeatedly shed your pack and dig out your scope every time a likely-looking spot appears. In the past I didn't carry binoculars on sheep hunts because of their extra weight. Now I usually carry light-weight, compact binoculars. Mine weigh in at only **eight ounces**. The convenience and ability to quickly scan a hillside is often worth their extra weight.

A high-quality, waterproof, light-weight spotting scope is a must for serious sheep hunters. You need to be able to look over long distances and identify sheep as well as determine horn size. The scope has to withstand wet weather and not fog inside. It should be a variable scope so it can be turned down in poor light conditions and when heat waves distort objects. I like a 15x-45x magnification range. A 15x setting is low enough for most circumstances and you can seldom use anything above 45x. Because of heat waves and other light limitations, 30x is the most common setting I use. There are a few times when 60x would be useful, but not enough to make me carry the extra weight of a larger scope. Look for a scope that weighs less than two pounds. Mine weighs 22 ounces.

photo by Kelly Stevenson

Will Stewart's impressive ram. High-quality optics are necessary to identify legal rams. Don't pass up rams like this one because poor optics hindered judgment. 36 4/8 x 14 2/8. Score—170 B&C.

    I also carry a light-weight (16 ounces), adjustable tripod to set the scope on. You can get spotting scopes with 90° eyepieces and not have to carry a tripod, but they are heavier and you expose yourself more when using them. You often need to just poke the scope over a ridge to stay low and out of sight. You can't do this as well with a 90° eyepiece. A tripod also allows you to sit up comfortably during long spotting sessions. You can take better pictures with it, too. Due to windy conditions and weight considerations I suggest a tripod that adjusts up to no more than three feet. That is all you typically need or can use in the mountains anyway.

Chapter 6

FOOD FOR THE HIGH COUNTRY

Sheep hunters need high-quality nourishment to meet the physical challenges of the hunt. When planning the food for a sheep hunt you need to consider weight, nutrition, palatability, bulk, and packaging. Preparation time and difficulty as well as the amount of water needed should also be considered. Since all of the food will be carried on your back, weight is a prime concern. Look for foods that have very little water content so they are concentrated sources of energy. Water is the heaviest component of most foods and you don't need to carry it unnecessarily However, don't select food based on weight alone. You should also consider your body's nutritional requirements.

Nutritional Needs

Your nutritional needs can be determined by looking at a table of Recommended Daily Dietary Allowances (RDA's). This table is published by the U.S. National Academy of Sciences and lists the RDA's for selected nutrients to maintain good health. The table establishes the nutritional requirements of a person based on age, sex, and weight. You should examine an RDA table and get a basic understanding of what your personal requirements are. In my case, as a 160 lb., 37 year-old male, I need approximately 2,700 calories and 56 grams of protein a day (as well as a variety of vitamins and minerals). However, these values are based on a moderately-active day. Strenuous activity increases these values dramatically. On a sheep hunt I burn up to 7,000 calories a day and my protein needs are also increased (as are my vitamin and mineral needs). This high caloric requirement is why I sometimes lose up to a pound of bodyweight per day on a tough hunt. A pound of bodyweight provides from 2,700 to 3,600 calories. The value depends on the relative amounts of muscle and fat burned to provide the calories. Knowing all this helps determine how to plan the food to take on a hunt. Because weight is such an important consideration, I seldom take enough food to provide 7,000 calories a day. However, I do look very closely at the nutritional analyses of the foods I take to get the most nutrition per ounce.

Calories Per Ounce

Calories per ounce (c/o) is the first value to consider when choosing foods for a sheep hunt. Most foods have a nutritional analysis printed on the package. By dividing the calories per serving by the ounces of food per serving you can determine the c/o. I generally don't consider a food for a sheep hunt if it doesn't contain at least 100 c/o. However, the amount of protein is also an important consideration. The following table illustrates how calories and protein compare between some foods.

FOOD	CALORIES PER OUNCE	GRAMS OF PROTEIN PER OUNCE
raisins	83	1
Pemmican bar	112	5
granola	125	2
chocolate bar	130	1
oriental noodles	133	3
dried milk	110	10
dried parmesian cheese	129	12
cashews	163	4
freeze-dried chili dinner	160	8
freeze-dried chicken stew	130	5
freeze-dried cheese omelette	150	11
canned chicken	32	7

All the items in this list meet my minimum energy requirement of 100 c/o except the raisins and the canned chicken. I have taken all the other items on sheep hunts. Also notice how much the protein values vary. I consider this value just as important as the caloric content. During a long day of sheep hunting you will break down a lot of muscle cells. You need an adequate amount of protein to rebuild them. You must take good protein sources as well as energy sources on a sheep hunt to continue to operate at peak efficiency. Some sheep hunters try to use a hunt as an opportunity to lose weight and, thus, don't take much food. That is a big mistake. You should plan on losing weight during physical conditioning if you need to. Don't jeopardize your sheep hunt by lacking the necessary energy to hunt at your maximum level.

Al Fedorenko with a full-curl ram (37 1/2") from the Tok Management drawing—permit Area. Hunters traversing the vast, rugged country where Dall sheep live can burn up to 7,000 calories per day.

Two other considerations when choosing food items are preparation time and convenience. Because of the time involved, I seldom cook breakfast or lunch on a sheep hunt. When I do cook, I want to do it as quickly and easily as possible. Most freeze-dried meals meet both requisites. They are prepared just by adding boiling water and letting them set for a few minutes. Additionally, most brands come in an inner package that can be used for preparation and serving. Thus, clean-up is minimal. Avoid foods requiring long cooking times—this requires you to carry more stove fuel. Also avoid foods needing large volumes of water for preparation. Although water is typically available, it is inconvenient to carry large amounts with the small containers used on a sheep hunt. Look for foods that can be prepared in one container. Then there is only one pot to carry **and** clean.

Meal preparation on a sheep hunt should be quick and simple. A stove, lighter, pot, cup, and spoon are all the cooking/eating gear a hunter needs.

Other important aspects of food items are palatability, bulk, and packaging. Suit **your** tastes when choosing food. Try out foods **before** taking them on a hunt. This is essential with freeze-dried meals. Some of them are **terrible**. But also realize that when you are famished after a long, hard day of hunting, **almost** anything tastes good. It is hard to successfully backpack soft, bulky foods like bread or fruit. Fortunately, most foods that satisfy the minimum c/o requirement aren't bulky. Packaging food properly is important, too. You can often discard the original wrappers and combine foods in one large, watertight, plastic bag. You can even discard the outer foil wrapper of freeze-dried meals—if you are sure you will eat them soon. Every fraction of an ounce counts when backpacking.

Freeze-Dried Foods

Freeze-dried foods have several qualities that make them a good choice for backpack hunts. They have a high calorie and protein content, they are light-weight, compact, easy to prepare, come in a watertight container, and offer a great variety of choices. You can buy freeze-dried breakfasts, lunches, dinners, snacks, and desserts. I use them often, but not exclusively. They also have drawbacks.

Though freeze-dried meals have a high c/o ratio, preparation and packaging also have to be considered. It takes about 1/2 ounce of fuel (with my stove) to prepare a dinner or breakfast; and the package weighs another 1/2 ounce—unless you discard the outer package which weighs 1/4 ounce. Taking this extra weight into account makes the high c/o ratio a little less attractive. Palatability of freeze-dried meals is also debatable, but there a few good ones I have found. I am not convinced these meals ever really become completely re-hydrated, either. Even though I use extra water and let them set longer than recommended, I don't think freeze-dried foods ever become quite "real." At least my system doesn't think so. Even one freeze-dried meal a day often has my digestive tract in an uproar. So, I use them sparingly.

If you have never used freeze-dried foods there is one more factor to be aware of. The meals labeled as two-person meals are **almost** enough for one person. The two-person dinners typically have about 700 calories in the entire package. That is not nearly enough energy for a hard-working sheep hunter. I often eat an energy bar or two with dinner as well as dessert which adds another 700 to 800 calories. Sometimes this total is still not enough, but it is closer to what I need to maintain my bodyweight. Plan accordingly for your body's needs.

Pasta

One item that I discovered years ago is a semi-cooked pasta called Couscous. It is a European product made from high-protein wheat. It has 100 calories and 5 grams of protein per ounce. It looks like little round pearls of wheat and cooks in five minutes when put into hot water. I always add some to my freeze-dried dinners along with more water. This makes a bigger and more reasonable dinner for me—as well as my digestive tract. Bought in bulk, it costs about **one-fourth** of what freeze-dried dinners cost. The dry Couscous is compact, durable, and easy to repackage. It is also versatile. I take spices, dried Parmesan cheese (a good protein and energy food), and dried vegetables to make a complete, tasty dinner. It is even better if you have some fresh meat to add. Dried vegetables are available from health food sections at supermarkets or health food stores. They are not freeze-dried, just dried, and my system likes them much better than the freeze-dried variety. Couscous can also be eaten as a breakfast grain just like oatmeal. The only criticisms of Couscous are it has only 100 c/o and may need to be prepared in a separate container. I never go without it, though.

Other options for a cooked meal are oriental noodles, various meal-in-a-cup lunches, and dried soups. The dried soups and cup-of-noodles are light and inexpensive, but they don't provide much nutrition for their weight. The noodles often come in a bulky cup that makes them even less attractive for backpacking. The oriental noodle varieties are inexpensive, a good source of calories and protein, and fairly easy to prepare (though they have to be boiled for several minutes). Although they are a little bulky and have flimsy wrappers that need to be replaced before backpacking, I use them occasionally. I often take out the little flavor packages and use these with my Couscous instead. This is a very convenient way to carry spices on a backpack hunt.

Ready-To-Eat-Foods

Most of the food eaten on a sheep hunt will be ready to consume right out of the package. Ready-to-eat (rte) foods save time and require no fuel to prepare. This is the most efficient way to carry nourishment on a backpack hunt. On very short hunts I sometimes leave my stove, pot, cup, and fuel behind (saving about 3 pounds) and just take this type of food. I don't recommend this for more than a short hunt as it is hard on your system and detracts from the hunt. There is something about cooking a meal that adds to the ambience of a hunt.

The author's fourth archery ram taken after an eight-hour stalk. The long, hard days common during sheep hunts call for plenty of nutritious ready-to-eat foods. 35 1/8 x 13. Score—147 0/8 P&Y.

Most of the rte foods I take are in bar form. The standard chocolate bars are okay, but they usually lack protein and good nutrition. I take them because they provide quick energy and are tasty. I also take granola bars which provide better nutrition, though they are not as tasty. I have discovered the best bars are a gorp-like bar called Pemmican bars. They have a good c/o ratio, a lot of protein, and other necessary nutrients. I often have one of these for breakfast before heading out in the morning. One of these bars quickly provides enough energy to get me through until lunchtime. Protein bars found in most supermarkets are also high on my list. They also have high c/o ratios and large amounts of protein—nutrition you need to continue climbing day after day.

Other popular rte foods are nuts, trail mixes, and dried fruit.

Nuts have very high c/o ratios, substantial amounts of protein, and they are compact. Trail mixes come in many varieties, but their c/o ratio is usually lower than plain nuts. This is because trail mixes have dried fruit in them which doesn't have a very high c/o ratio. I often take dried fruit, though, because it provides quick energy and some bulk, which backpacking foods often lack.

Jerky is also an excellent rte food for sheep hunters. It is very high in protein and has a high c/o ratio. Wild game jerky can be made easily at home with a standard oven on a low setting. I marinade strips of game meat in a teriyaki sauce first and then dry it for 10 to 20 hours—depending on how thick it is cut. Hamburger jerky can also be made at home by rolling ground meat into flat layers on waxed paper and then drying it at room temperature. After cutting it into strips, it stays together just like regular jerky—and it has a very high c/o ratio because of the added fat. It doesn't keep as well because of this fat, though, so it should be used quickly if not refrigerated.

Backpackers can also take instant puddings and other desserts. These are sort of a luxury, but they do have a high c/o ratio. The instant varieties only need cold water and a little stirring to prepare. I also take the filling package of an instant cheesecake and add fresh blueberries for a mountainside dessert. These can be unbelievably delicious to famished sheep hunters. If desserts need milk to prepare, just premix the proper amount of dried milk with them and repackage the result.

Although bread is not a common backpacking food, I sometimes take it to add bulk to my mountain diet. Bagels and unleavened breads are suitable for backpacking. They are durable and not too heavy. I sometimes take these and plan on either eating them in the first few days or leaving them at base camp. You can even take peanut butter (good c/o ratio and protein source) to go with the bread. Remember to take necessary precautions since wild animals love peanut butter.

One item I **always** take on a sheep hunt is powdered teriyaki marinade. This is an excellent way to prepare any wild game—particularly sheep meat. After a few hours in this marinade, meat roasted over an open fire will be some of the **finest** you have ever tasted. I have not eaten any tastier meat than wild sheep prepared this way!

Drinks

As I pointed out earlier, it is **very** important to drink large amounts of liquid while sheep hunting. Hunters can also get some **energy** from fluids with powdered drink mixes. I often take powdered fruit juices or electrolyte mixes and dilute them to half strength. It is a real boost to have a quick-energy drink while climbing steep hillsides. I dilute them to half strength because I drink such large quantities and want to make it last longer. Don't try to save on weight by taking mixes with a sugar substitute. They are fine if you just want the flavor, but they will not give you any energy because they lack calories. The sports-drink mixes are also good choices because they provide electrolytes that we need to replace. Salt tablets may be needed if you are sweating profusely. However, freeze-dried meals and other packaged foods have always provided me with **plenty** of salt.

Other drinks you can take are dried milk and hot mixes. I often take dried milk. It is a good energy and protein source and it adds variety. Hot cocoa or coffee can be taken for cold mornings. Cocoa is basically sugar so it provides energy, but coffee (black) has few calories. Also, drinks with caffeine (a diarrhetic) deplete us of fluid rather than replace it. Consequently, I suggest avoiding coffee **and** cocoa when water is scarce. I often drink hot fruit juices instead since I take it along anyway. This way I take in energy and fluids as well as warming up on cool mornings.

Wild Foods

One attractive option for a backpack hunt is to utilize wild foods to supplement our diet. I have explored this possibility and concluded it has limited use for sheep hunting. There are several handbooks available that identify wild, edible plants in Alaska. However, most of these plants are found at lower elevations than you will frequent when sheep hunting. The few I have been able to find at sheep-hunting elevations are available predominantly in the spring or early summer. **Sometimes** you can find young fireweed or other plants on the edges of snowfields in the fall that are still edible—mature plants are often too tough to eat. Of course, at times you will be traveling at lower elevations where more plants are available. You cannot count on these, though, and most have few calories and little protein. I don't pass them up, however, because they do have vitamins, minerals, and roughage I need in my diet. Mushrooms can be found even at high altitudes late in the year, but the edible varieties are difficult to identify and they provide almost no food value of any kind.

Few wild foods are available in sheep country.

The only wild foods sheep hunters frequently consume are berries. There are several varieties available and they are easily identified. Of course you need to educate yourself so you can avoid the few that are poisonous. Berries often blanket upper hillsides. They are a great supplement to our diet. Hunters can easily stop and gather a quantity for a later meal (cheesecake?) or just gorge themselves on the spot. Berries are excellent sources of energy and fluids. Berries also provide vitamins, minerals, and roughage needed during a sheep hunt.

Of course there are small game animals and sometimes fish available on a sheep hunt. I have experimented with rat traps for ground squirrels and shotshells for my rifle to try to take advantage of these protein sources. My conclusion is that it takes too much time and effort to do this. Ground squirrels in Alaska are **small** animals with very little meat; game birds are often difficult to bag. I can't justify the noise and effort necessary to take them. If they are **easily** available, I will take game birds, but I don't **count** on them.

Seldom am I close enough to a fishable body of water to even consider taking along fishing gear. The only fresh meat I count on is **mutton**.

Supplements

I think taking vitamin tablets along on a sheep hunt is justified. The hunt itself is physically demanding. During a sheep hunt we put in long, hard days and may not get much sleep. Plus, we have a limited diet that is short on fresh fruits and vegetables. This creates a situation where we are probably not getting enough vitamins, minerals, **or** calories. Vitamin supplements weigh very little, so the only consideration on a backpack hunt is their cost. I think it is worth the minor cost to include them. They can only improve our ability to keep going strong on a long hunt. Perhaps they will help you go that extra mile that is often necessary.

I also take protein powder along. I mix it with a juice mix or powdered milk. It is light-weight and worthwhile insurance that I am getting enough protein.

Packaging

Packaging food properly means discarding unnecessary wrappers (weight) and repacking in waterproof containers. Most food items come in the wrong size, shape, or type of wrappers to be taken as is. Go through your food carefully and package items in as few containers as possible to save space and weight. I often use Ziploc bags. They are light-weight and watertight. I can pack all the Couscous or drink mix I need for an entire trip in one package this way. As I mentioned, you can discard the outer foil wrapper from freeze-dried foods and repack several meals together. You will have less bulk to pack in **and** less garbage to carry out if you repackage properly.

There are all sorts of plastic bottles and containers available to carry foods in while backpacking. Check the weights of these containers before you take them along, though. They are surprisingly heavy. I only have one or two I take and they are used for good reasons. Once I get everything repackaged, I put everything in a light-weight nylon bag to further protect it and organize my pack.

Food is **heavy**. It should be packed fairly low and near the front of the pack to balance the load properly. You should allow at least 1 1/2 pounds of food per day per person on a backpack hunt. Even if you choose foods that have great c/o ratios, this is only going to provide about 4,000 calories per day. It would be prudent to take more food in case you stay out longer than expected. In Alaska, planes don't always come when scheduled. So taking extra food for this contingency is recommended. You can always eat the extra food once you **know** the trip has been shortened or if you get a fresh supply of meat. Even if you don't take this extra food along in your pack, it can be left at the airstrip so it is available if needed.

*Sheep hunters must prepare adequately for a hunt. This is no time to have doubts about yourself **or** your gear.*

FOOD FOR THE HIGH COUNTRY

I feel preparation is **80%** of a hunt—and choosing the right food is a vital element. Carefully select the food for a sheep hunt. It is the last phase of Part I—Preparation. Next comes **Part II—Hunting.**

PART TWO- HUNTING

Chapter 7

SHEEP BEHAVIOR

I **thoroughly enjoy watching sheep.** One early-season hunt immediately comes to mind. Opening day of sheep season several years ago found my hunting buddy and I poised for the final stalk. As we awaited the right opportunity, the largest ram among the two dozen or so spread out in the cirque above us put on a unique display of social dominance. A driving urge to exhibit his dominance compelled him to charge across the mountainside challenging other rams. He sprinted from group to group and attempted to butt heads with any willing ram. Unfortunately, there were no takers this early in the season. They all just moved away and ignored his challenge. Undaunted and feeling unbeatable, he would sprint to the next group of rams and repeat his challenge. This is by far the most dramatic dominance display I have witnessed among Dall sheep, but it is only one of many fascinating sheep behaviors.

An important element of the sheep hunting experience is becoming part of their lives for a brief period. We get up with sheep—watch their feeding, resting, and social behaviors—and often put them to bed—all in the same day. An understanding of sheep behavior and biology increases the pleasure of hunting them. It also improves our chances of success. We know where and when to look for legal rams, which helps us predict their behavior.

Population Dynamics

The last population estimate for Dall sheep in Alaska by the Alaska Department of Fish and Game (ADF&G) was 70,000-75,000 animals in 1983. Since then the population has declined to 50,000-60,000 animals and it may still be declining (Heimer, 1993).

Dall sheep populations generally exhibit stable population levels over long time periods; population growth is slow and limited by suitable habitat (Heimer, 1988). "Wind action, snow depth, and snow hardness appear to be limiting factors in determination of suitable habitat. Prevailing winds are required to reliably remove snow from winter food." (Heimer, 1988). Thus, some apparently suitable habitat in Alaska is devoid of sheep because of incompatible snow conditions.

Factors which can reduce sheep populations significantly are weather, disease, and predation. "The greatest weather problem (at least in frequency) is deep snows which are associated with poor lamb production the next year, probably through a nutritional problem of limiting ewe access to higher-quality food." (Wayne Heimer, 1994). Crusty snow conditions caused by a winter thaw/freeze cycle also impair feeding efficiency. Both conditions increase winterkill. Additionally, animals weakened by poor feeding conditions are more susceptible to disease and predation.

The obliteration of numerous wild sheep populations caused by the introduction of disease(s) from domestic sheep is well-documented among bighorn sheep. Although Alaska sheep populations have so far been spared this type of plague, recent work at Washington State University has shown Dall sheep are much more susceptible to Pasteurella pneumonia than bighorns (Wayne Heimer, 1994). Dall sheep populations must be protected from this very **real** possibility. In fact, just a few years ago a domestic herd of sheep was grazing on the lower reaches of the Chugach Mountains' western edge. This area overlapped Dall sheep habitat. Luckily, the domestic sheep have since been removed. Effective teamwork between a few diligent state biologists and the Alaska Chapter of the Foundation for North American Wild Sheep (FNAWS) resulted in prudent testing of members of the domestic herd for disease—which proved negative to everyone's relief. However, the current devastation of some herds of the Brooks Range sheep population is still being analyzed by state biologists. Although they know disease is the cause of the die-off, they have not conclusively identified the disease(s).

Predation can also reduce sheep populations significantly. Particularly wolves, and maybe coyotes and eagles can depress local populations (Heimer, 1988). Dall sheep are particularly susceptible to predation in the spring when they come down to feed on the year's first green growth. When they are at lower elevations, away from escape terrain, predators are much more successful. Sheep certainly realize the danger, but they cannot resist the green food after a long winter of dry grass and browse.

For a few weeks in spring, rams can be seen near timberline feeding on the first new grass. These seven rams are bedded among the alders in early May.

Incidentally, spring is also a great time to observe and photograph sheep because of their accessibility. Large rams who live alone in remote glacial badlands are sometimes reachable for a brief time during spring. Higher concentrations of sheep also make this a great time for observing social interaction.

Social Behavior Patterns

Sheep have well-developed social systems (Heimer, 1984). This is quite obvious to anyone who has spent even a short time observing Dall sheep, especially hunters who closely scrutinize every movement of their quarry. As sheep hunters we should understand the basics of their social system because mature rams—our quarry—are necessary for it's maintenance. Knowledgeable sheep hunters can also understand management decisions and are more likely to make hunting decisions which have a positive effect on sheep populations in general.

Socially-mature rams improve the social order of Dall sheep populations, but not without a cost. The cost is increased mortality among mature rams. One study (Heimer, 1986) found that ram mortality was 40%-60% up until one year of age, it dropped dramatically to 2.3% from one to eight years, and then increased again to 17.8% after eight years. Geist (1971) reported similar findings. Eight years seems to be the average age at which Dall rams first become fully-involved in the rut. The rut occurs in November and early December in Alaska. Because rutting depletes energy reserves just before winter, mature rams exhibit a significantly higher mortality than immature rams.

This winter-kill has eight growth rings. Ram mortality increases significantly after eight years. 38 x 14. Score—160 B&C.

From a hunter's perspective this means we should never harvest all the mature rams from a herd. Some mature rams **will** winterkill, but this is unavoidable if we want to maintain healthy, well-ordered sheep populations. This should be our overriding goal anyway.

The methods sheep use to maintain social order are intricate and fascinating. Whenever I watch sheep I always observe social interaction. When there are two or more sheep present, you can bet they will act according to a predictable social structure. In his book *Mountain Sheep and Man in the Northern Wilds* (1975), Valerius Geist explains the precise dominance order that mountain sheep develop. He says each ram within a group of rams knows his individual place in this order; there is a 1,2,3, etc. dominance relationship between rams and it is strictly adhered to. This relationship is maintained year-round by dominance displays and physical reinforcements (butting, kicking, hooking, etc.). Ewes and young sheep also display dominance behavior in the same manner as rams, but since they are not our quarry, I will only mention them from now on to explain the behavior of rams.

The place an individual ram occupies in this dominance order is dependent on his horn mass. Geist explains exactly how this occurs in another book, *Mountain Sheep-A Study in Behavior and Evolution* (1971). Geist says immediately after butting heads, rams display their horns to each other. Presenting, twisting, and the low stretch are common postures rams use to display their horn mass from all angles. Because of this thorough display, rams are able to relate the force of the blow they just received to the horn mass of their opponent. Thus, rams learn to accurately judge their relative social position within a group of rams just by looking at horn mass. This allows social structure to be established and maintained with less physical confrontation.

Although horn butting occurs all year long, it is more prevalent just before the rut. During this period rams travel long distances, thus, meetings between strange rams occur more frequently. However, because of their ability to judge dominance by horn mass, rams don't butt heads with all the strange rams they meet. Rams of equal size still have to actually butt to establish dominance prior to the upcoming rut, but excessive butting is avoided because of this judging behavior (Geist, 1971).

The effects this dominance order has on behavior can be

astonishing. While photographing sheep one summer in the Wrangell mountains of Alaska, I observed two rams who demonstrated the great reliance sheep place on this aspect of their social structure. One ram was obviously larger than the other and the smaller ram was his subordinate. I crawled within 70 yards of them and let the number two ram see me. I was in a dip in the tundra which hid me from view of the number one ram. For the next 30 minutes the small ram nervously looked between me and the large ram as he tried to decide why they were not fleeing this apparent danger. Finally, the subordinate ram went back to feeding since his leader indicated (by his actions) that everything was OK. Once I let the number one ram see me, though, they were off and running. I could almost see the scowl on the face of the smaller ram as he lost faith in the wisdom of his social superior.

Grouping of sheep during the hunting season is fairly uniform, with some exceptions. Mature rams tend to run in bands of one to forty, with two to eight being the average. Ewes, lambs, and young rams (yearlings and two-year old rams) are grouped together in bands up to 100 animals. Frequently two equal-size rams will run together, but often a large ram will run with a much smaller ram. The one difference between these pairings I have observed is that the equal-size rams will usually stay together, while a large ram will often desert a much smaller buddy. Occasionally, you will see rams as large as three-quarter curl running with ewe bands and, although rare, even legal (full-curl) rams are sometimes spotted among ewes. On one hunt I found twenty rams (half were legal) on the same mountainside as eighty ewes and lambs. Although the rams were grouped by themselves in fours and fives, they occupied the same meadows and bedding areas as the ewes for three days. I later learned this was a common occurrence in this area.

Single sheep are also commonly seen. Old rams often prefer to be by themselves. Also, yearling sheep frequently explore on their own—like all youngsters are prone to do.

Rams tend to be nearer the heads of drainages in rough country, while ewes and their young tend to be lower in drainages on gentler slopes. Ewes and lambs are fully capable of traveling the crags and cliffs that rams live in, but just prefer the better grazing available at lower elevations. The elevation at which you will commonly find rams in Alaska's mountain ranges varies from 2,500 ft. to 8,000 ft.

Generally, look for the highest, roughest peaks to find rams. If you are spotting lots of ewes and lambs, but no rams, look higher. Although there are no absolutes, rams tend to be many miles distance from large bands of ewes and lambs during hunting season.

Defenses

With a few exceptions—like the Tanana-Yukon Uplands—Dall sheep live in open, treeless country in Alaska. They are visually-oriented animals; they depend on their eyes to warn them of danger. They also watch each other closely. The alarm posture for sheep is similar to other big-game animals; their head and neck are erect and they walk with tense steps (Geist, 1971). The use of others within the flock to warn them of danger in this way as well as what they see themselves has protected the species through time from their main predator—wolves. Their basic strategy is to stay in high, open terrain where any danger can be spotted at a great distance. Once danger is spotted, they use cliffs and rocky slopes to escape. Their surefootedness in the cliffs and crags allows them to travel where neither man nor beast can follow.

Sheep have very good eyesight and adults are looking for danger about 80 percent of the time. Single sheep and older rams are even more nervous and seem to be watching for predators 90 percent of the time! I do not think I am exaggerating their alertness one bit—they are very vigilant animals. I have had a ram spot me at over a mile and promptly leave the country. This was after getting just a two-second glimpse of only a fraction of me on a ridge. I watched that ram travel two miles and 3,000 vertical feet before going out of sight at a trot. When spooked like this, sheep often travel many miles and several ridges before stopping. I often watch sheep just at dusk and they will look hard until it is too dark to see. They want to make sure nothing is approaching before they commit to a bed for the night. Once bedded, sheep will stay in one spot all night—unless disturbed. As the light of morning fills the mountains they will be watching intently to check for any danger before they get up to feed. Whenever they do see something suspicious, they stare a lot. I have had them stare at me for 90 minutes trying to figure out if I was a threat. If you want to hunt sheep successfully, you should realize how **much** they rely on their keen eyesight.

Sheep seem to have a fairly good sense of smell, although they cannot **rely** on scent to alert them to danger in the mountains. Wind patterns in the mountains are very erratic. There are predictable down-slope and up-slope movements of air in the morning and afternoon, respectively, but the uneven terrain affects **surface** air flow significantly. The wind frequently swirls and dances in tight pockets of irregular, rocky structures. The wind also commonly lifts scents off the surface and disperses them at high altitudes without ever touching down as it would on gentler slopes. I have often been directly upwind of sheep without alarming them.

However, you cannot ignore the wind. If sheep scent you, they will spook just as fast and far as when they see you on a skyline. A few years ago I had two stalks ruined in one day because the wind abruptly changed direction 180 degrees. Both times I had correctly anticipated the direction the rams were traveling and was committed to one spot. The rams got within 100 yards each time before scenting me, but this is still out of my range as a bowhunter.

Sheep also have good ears, but again, their habitat decreases the usefulness of sound as a reliable defense mechanism. Rocks are continually falling in typical sheep habitat. Sheep are accustomed to tumbling rocks and pay little attention to their sound. However, they **will** notice footfalls or suspicious rock slides caused by careless hunters. Also, any other strange sounds will certainly alert them. The level of noise you can safely make during a stalk is just a little bit higher with sheep than most other big-game animals.

Sheep tend to indicate their anticipated direction of travel—to an alert hunter. They will often look for hours or even days in one direction before heading that way (Geist, 1971). I once watched seven rams poised at the upper edge of a timbered hillside overlooking a deep river canyon and stare downhill for an hour or more. When they were satisfied it was safe, they headed toward a well-used crossing 1,500 feet below timberline. This habit can sometimes alert perceptive hunters about an upcoming movement of their quarry.

Feeding Habits

Dall sheep feed primarily on grasses and sedges (Heimer, 1984) and they have two primary feeding periods during the day. Typically, rams will feed for two to four hours in the morning and two to four hours in the evening. Just as soon as visibility is good enough for sheep to see any obvious danger, they rise to feed for a few hours in the morning. The evening feeding period also lasts just a few hours, and ends before dusk limits visibility. While feeding, rams look a **lot**. Typically, a ram will graze for 10 seconds or so, then lift his head and scan the area for 30 seconds. Rams actually do **more looking than feeding.**

The timing of these feeding periods varies greatly as the day length varies during the hunting season. This timing also varies with the latitude of the hunting area. In the far north where there is perpetual daylight early in the season, sheep still bed down for the "night." The basic pattern is the same, only the times change. Often, rams will have short, midday feeding periods. These seem to be more of a way to stretch, though, as midday feeding is not usually consistent nor very long.

Rams will often descend 2,000 to 3,000 feet to reach good feeding pastures. After a few hours of feeding, they will ascend back to their bedding area until the next feeding time. Rams will often follow such a pattern for many days if undisturbed. Frequently, their bedding area in such cases is unreachable by hunters. The only way to hunt them is by identifying this feeding pattern and laying a trap along their route.

Sheep sometimes just travel **horizontally** to feeding grounds. Mature rams in particular like to bed in the rocks, whether alerted or not. They will then move as little as possible to find good grass. Even the seemingly barren cliffs rams like to frequent are not devoid of food. Close scrutiny will reveal there is often sufficient grass to sustain rams for weeks. Dall sheep don't water, so they have no other need to leave the crags and cliffs if there is enough food present. The little patches of bunch grass sheep like are frequently not visible from below. Hunters who typically glass sheep from below often think there is no food on the cliff faces. They wait days in the belief a particular ram will have to move to get food. That is seldom the case, and these hunters may never understand what went wrong with their plan.

Sheep frequent mineral licks and have favorite spots where they even **eat** the soil for the high mineral content. Sodium and magnesium are the most likely target minerals (Wayne Heimer, 1994). They will often travel long distances to get these minerals. Large groups of ewes as well as rams can be regularly observed at these locations. It is postulated that sheep need to replenish their bodies' minerals after a long winter of poor quality feed (Geist, 1971). This is predominantly a spring and early summer occurrence. It seldom continues into the fall hunting season, so it is of little use for hunting strategies.

Resting Habits

Dall sheep are ruminants and they spend much of the day chewing their cud. During the hunting season, undisturbed rams will be found resting from mid-morning to mid-afternoon. After their morning feeding, rams will bed where they can see approaching danger. The bedding area will either be in or near escape terrain. Each ram in the group will lie facing a different direction. They will watch each other to see which direction others are facing and then pick a different one for themselves. As a hunter, I prefer to think this is so they can watch as many directions as possible. Behavioral scientists attribute this to social interaction—they are trying to avoid eye contact with other rams (Geist, 1971). Regardless of the reason, the result is the same—rams within a group will **all be facing different directions.**

This resting period is fairly uniform, but there may be some slight interruptions. Once rams bed for the day, you can usually count on them to be in the same approximate area for several hours. However, they do frequently get up to stretch and change positions—about once every hour or so. They may also get up and feed for short periods at unpredictable times. Younger rams tend to do this more than older rams. You will also see midday feeding more often among large groups of rams than groups of two or three. The length and consistency of resting periods is also somewhat dependent on the quality of feed. If rams are undisturbed and have access to high-quality feed, they will more often bed for the entire period, except for stretching. If they are living among the rocky faces with little grass—typical habitat of larger rams—they may be feeding at any time of the day.

After this detailed explanation about predictability, I want to talk briefly about unpredictability. Over the past few seasons I have noticed that rams seem to bed for shorter periods of time. Stalking undisturbed rams used to be simplified by the assumption they would stay bedded for several hours in the same small area. However, rams seem to be moving more during the day lately. I haven't concluded whether rams are adapting to hunting pressure to survive or I don't know as much about them as I thought.

*This **large** ram was traveling in the middle of the day for unknown reasons. George Banks (pictured) was guided by the author. 41 x 14 1/8. Score—168 4/8 B&C.*

The location of a ram's bedding area varies. Frequently, undisturbed rams will bed just at the top of a grassy meadow several hundred yards from escape cover. Older and wiser rams usually choose to bed closer to escape cover or actually in the unapproachable cliffs and rocky chutes. This is particularly true during hunting season after rams have seen hunters. They may then locate beds on small ledges just big enough for one ram. Beds of this type are found throughout sheep country and are used year after year. They often provide a ram with a grand view of an entire drainage while still concealing him from hunters below. Older, wiser rams seem to seek out these types of beds more frequently. Often only their head is visible. The average hunter will seldom spot these rams. Rams also seek out shady beds when temperatures get too high. Even a white sheep is hard to spot when lying in the shade.

A ram's grassy bed. This vantage point allows a ram to see anything approaching from below.

Seasonal Movements

Dall sheep have summer and winter ranges that may be separated by many miles or be fairly close together. The distance between the two is determined by the nearest reliable winter range with high-quality forage (Wayne Heimer, 1994). The timing of their seasonal movements between the two is somewhat weather-dependent. In most sheep country there are well-worn trails connecting summer and winter ranges. As the snow melts back toward the upper reaches, sheep begin to follow the fringe of new grass as it sprouts. Long lines of rams can often be seen during spring as they slowly make their way back to summer feeding grounds. At this time of year, rams will often be far away from the secluded backcountry habitat they prefer and at lower elevations than any other time of the year.

The movement back into the tops of drainages and the glaciated areas many rams prefer takes weeks and even months in some cases. Snow melts slowly and new growth comes up even slower. I have seen rams in the middle of July that were still heading back toward the upper reaches of their habitat. The extent to which they travel depends on how far back food is available each summer. I am convinced that as the hunting season begins, some reclusive rams are still moving back toward more remote feeding grounds. These rams basically keep moving until new snow forces them to turn back toward wintering grounds.

As hunting season begins in Alaska, snow is just beginning to fall at the upper elevations of sheep habitat. The start of the major fall sheep migrations usually coincides with the last few weeks of hunting season. However, ram movements are weather-related. A significant snowfall (over two inches) that covers feeding and bedding grounds will usually move rams down and out from their summer range. This movement is usually just temporary as they will follow the melting snow back up in a few days. Therefore, undisturbed rams are frequently found on the south-facing slopes at the lower edge of the snow line in the fall. They prefer the drier bedding areas and the uncovered grass available on the sunny sides of mountains. Rams **usually** just move up and down in the late fall with the fluctuating snow line but they may start their migrations early if the snow is heavy enough. I know hunters who've seen every ram for miles move out of an area because of an early, heavy snow—and **not return** that fall.

The largest ram I have taken was pushed out of glacial backcountry by the first heavy snows of fall. I located 40 rams in one basin where they had gathered together from their individual drainages. Ninety percent of them were legal rams brought together by early snows at higher elevations. The ram I took generally lived in remote, inaccessible habitat with little hunting pressure. The snow not only brought him out, it also brought him down from the cliffs that had protected him.

Rutting among Dall sheep occurs in late November and early December (Heimer, 1984), which is **after** the hunting season in Alaska. The fall migrations precipitated by snow and colder weather gathers sheep of both sexes for the rut. Although this may not affect hunting strategy because of the timing, it is an interesting part of the life history of Dall sheep.

There are some noteworthy exceptions to these general habits I have described. As Geist says "It is quite possible that old and young sheep are less predictable in their movements than others." (1971). I have often observed—and frequently heard about—old rams which live by their own set of rules. These are rams which live in very isolated areas. They may not come out in the fall or participate in the rut. These are the rams trophy hunters are continually looking for.

The unpredictability of older, larger rams is also common knowledge among veteran sheep hunters. Rams who are unpredictable have an edge when it comes to surviving to trophy size. Recently, I hunted a very large ram who was running with a much smaller (38 inch) ram. We let him out of our sight for two hours and he suddenly disappeared. The smaller ram was still in place when we got within 70 yards of him, but the large one felt a wild urge and vanished. There was no reason for him to leave and he had to travel 2,000 feet uphill to be out of our sight. The larger ram didn't come back in the three days we stayed there, so his penchant for sudden moves had saved him; and I bet it was not the first time his erratic behavior foiled hunters. This type of behavior is one of the reasons trophy rams get that way.

Craig Wawers and his 40" ram. This large ram was living in a brushy river canyon. He was very difficult to spot from the air. 40 x 13 2/8.

Some sheep live in river canyons. One sheep I guided a hunter to was living in a brushy canyon at no more than 2,500 feet elevation—and he was a "forty-incher." Frequently sheep will live in these canyons at lower elevations if no one pressures them. They are usually relatively easy to harvest, so this is not as common in heavily-hunted areas. However, many hunters would be surprised if they took the time to look these areas over.

Caves are another unusual place to find sheep. I have only seen sheep in caves a few times myself. However, I often hear about cave sheep, so I always look them over carefully.

The author photographed these sheep in mid-summer as they moved freely in and out of this cave. Although sheep use caves only occasionally, hunters should always check them out.

Lifetime Habits, Movements

"all sheep are loyal to their home range and their social group." (Heimer, 1984). Weather permitting, mountain sheep will be back at their home range year after year (Geist, 1971). If rams cannot be found where they typically have been found each year, they are usually dead. However, before deciding this is the case, look **very** carefully. I've heard many stories about rams that miraculously appear at the end of the hunting season—after they were presumed dead.

Dall sheep have fairly predictable daily, seasonal, and lifetime patterns they follow. Understanding these patterns will help you hunt them as well as appreciate their life history. A large part of the satisfaction of hunting is getting to know our quarry. Each hunt adds to our knowledge and appreciation of sheep. In the next section, I will explain how you can use this understanding of sheep behavior to **hunt** Dall sheep successfully.

Chapter 8

HUNTING STRATEGIES

Strategically, Dall sheep are not particularly difficult animals to hunt. A hunter with a good understanding of sheep social behavior patterns and defenses has a good chance for success—assuming the hunter is in good condition, has the right gear, and there are legal sheep in the hunting area. Jack O'Connor also felt that [Dall] sheep were relatively simple to hunt. However, sheep hunting is a unique challenge and there are some basic principles to follow.

Basics for the Hunter

The following guidelines will improve your comfort, safety, and chances of success on a sheep hunt:

-Always keep **safety** foremost in your thoughts.

-The most dangerous element of sheep hunting is not falling, it is the danger of succumbing to **hypothermia**.

-Climbing up steep, rocky chutes or faces is easier than coming down. This means if you can just barely make it up, you should reconsider--if you have to come down by the same route.

-Try to stay **cool** while sheep hunting. Dress in layers so you can take off clothes as you get warmer. By staying cool, you will not sweat as much—thus your clothes will be drier. Dry clothes are more comfortable and more likely to keep you **alive**!

-Drink and eat often. Your safety and your abilities will be greatly enhanced if you are properly fueled and hydrated.

-Unless absolutely necessary **do not** use **all** your energy reserves. Hunting opportunities and dangerous situations are frequent during a sheep hunt and both call for sudden bursts of energy.

-Shiny hands and faces stand out for **miles**. Cover them with gloves, head nets, hats, and camouflage paint whenever possible.

-Cover shiny gear—packs, slings, watches, etc.—with paint or tape.

-**Don**'t **skyline yourself!** Make a habit of traveling just below ridgelines and peeking over periodically.

-Slowly peek over ridges, show as little of yourself as possible.

When looking for sheep, beware of their keen eyesight. This spotting scope is placed to allow a hunter to stay hidden while glassing for sheep.

-Always catch your breath before peeking over ridges. Be prepared for sudden activity if an opportunity presents itself.

-Always be alert for sheep and be ready to freeze if caught away from cover. Like most big-game animals, sheep rely on **movement** to alert them of danger. Often, even if a hunter is visible, but **stationary**, sheep will not spook.

HUNTING STRATEGIES

-Cross streams early in the day, **before** they swell from meltwater. Face upstream and plant the pole upstream so you can lean **into** the current.

-Remember that water drains off rocky watersheds much quicker and raises water levels much faster than green drainages—plan accordingly.

-Locate camp out of sight from hunting areas as much as possible. Sheep hunters are often pinned down in their tents when the morning light reveals unexpected sheep close to camp. Forethought can usually eliminate this possibility.

-Prepare to bivouac away from camp if necessary. I seldom leave camp without my sleeping bag as a safety precaution. On one hunt I had to bivouac two nights after a kill before I could get back to camp. If I had not taken my sleeping bag, my situation would have been very uncomfortable—**at the least!**

-Learn to distinguish rams from ewes at a distance by the angle of the neck and the stocky body. The weight of a ram's horns causes his neck to be more upright and over his body than that of a ewe. A brief glimpse at a distance is sometimes all you get of a sheep before it goes out of sight. Having this ability to distinguish rams quickly can save you a lot of time and effort or mean the difference between success and failure.

-**Have patience!** Sheep appear suddenly, weather changes just as suddenly, and fog can appear out of nowhere to facilitate stalking.

-Plan ahead before you shoot. You must be able to **recover** a sheep where you shoot it **or** where it may fall after the shot. Also, although sheep horns are tough, a long fall can break horn tips and/or destroy the cape as well as most of the edible meat.

Basics of the Hunt

Dall sheep live in open country and use their keen eyesight to spot predators. However, **we** are also very visual animals and have the advantage of **optics**. This makes our visual abilities superior to theirs. The first rule of sheep hunting is to see your quarry before he sees you. Sheep hunters typically spend **hours** behind a spotting scope every day of a hunt.

Patience is required while spotting for sheep. It is difficult to stay hidden in one spot for hours and glass seemingly barren mountainsides. The best time to spot rams is early morning and late afternoon when they are feeding and moving—typical strategy for most big-game hunting. However, since sheep live in open country where they should be visible, it is easy for hunters to give up looking too **soon**.

Hunters typically spend hours behind a spotting scope looking for sheep.

Sometimes just the slightest crease in the hillside or other break in the terrain will hide rams. I recall one hunt where I had been stalking six rams all day. I was trying to maneuver within bow range of any legal ram in the group. I had given up and discounted one four-foot boulder near the bottom of the valley. I thought it was much too low and vulnerable for any legal ram to hide behind. At dusk, as I hurriedly descended two thousand feet, he suddenly appeared at 20 yards from his bed—below the boulder. I was too slow nocking an arrow. He escaped and set a record time for crossing the valley. I have also failed to spot entire flocks lying in the shade, on snow fields, or at unexpectedly low elevations in the brush.

The distance a ram will run once he is spooked depends on the situation. The terrain, hunting pressure, the ram's size, how he was alerted, how many rams are in the band, and numerous other factors all affect how far he goes to escape. I have had one of the lesser rams in a band spot me from 100 yards. His nervousness moved the band **just** around a hillside. There have also been occasions when a lead ram has **briefly** glimpsed part of me from over a mile and led his band **several miles** away at a trot.

Mountain terrain is full of irregularities that effectively hide sheep. The distant hillside harbors over 100 sheep.

O'Connor said a spooked ram will leave the mountain, but he will usually come back. However, Geist (1975) gives an example where rams deserted part of their home range for the rest of the season after being spooked by hunters. There are no absolutes; each situation is unique.

Sheep are adaptable (Stapleton, 1991) and they are adapting to the considerable hunting pressure in Alaska. Generally, rams have encountered hunters from at least the time they become half-curl and join a band of larger rams. By the time they become legal, they have three or four years' experience with hunters. Dall sheep currently tend to live in rougher country because of hunting pressure. There is one well-known steep, nasty rock face in an area I hunt that I call "the wall." Sheep have learned that they are safe from hunters on this three-mile stretch of rock. There are always some very large sheep on this wall and, although hunters are commonly in sight, the sheep stay put all season. Occasionally, a hunter takes a risk and hunts "the wall." I did it once and was amazed at how relaxed the rams were. I watched one ram within 50 yards for 30 minutes and he never once looked up from his feeding. He was so confident his position was unapproachable by any predators, he failed to notice me even when I stood up on a rocky spine to shoot.

Some hunters claim they have successfully approached sheep by casually strolling toward them while in full view. Hunters risk spooking their quarry out of the country by doing this. If the distance is over one-half mile and the hunters **angle** toward the sheep it may work to reach necessary cover to complete the stalk. But Geist (1971) points out that sheep mentally anticipate where a predator should reappear when they angle out of sight. I don't recommend this strategy for Dall sheep.

Determining which ram to shoot among several in a band is difficult and often a split-second decision. The lead ram is usually the largest. That is, he will probably have the most horn **mass** and score the highest. The largest ram in a group is also commonly the wisest. The others will rely on his decision concerning danger. When slightly alerted, the lead ram will most likely be the one who stands and stares at the possible source of danger. Frequently, the other rams will go back to feeding until the lead ram decides whether to flee or stay. His decision can be affected by other sheep, too. Rams or even ewes up to a mile away who are obviously spooked will also cause a lead ram to run.

One surprising habit I have discovered is the tendency for sheep to relax once one individual of the group is down. On almost every occasion when we have made a clean kill on an unsuspecting ram, the others simply **walk** away. When I walked up to my first bowkill, his partner—who had forty-two inch horns—turned his back at 20 yards and just ambled away! They seem to sense when a predator has taken a sheep, the others are fairly safe. In the aforementioned situation, several legal rams watched me for two hours within 200 yards as I cleaned my sheep. This is typical of the one-shot kills I have witnessed and it is worth remembering when a hunting partner also wants a sheep.

HUNTING STRATEGIES

This ram was taken minutes earlier with one rifle shot. Fifty yards away are five mature rams. They are slowly walking away as the author approaches. Rams often become complacent toward hunters after one ram is dead. 36 x 13 1/8. Score—150 2/8 B&C.

Another habit to remember is the tendency of two large rams who have been running together to remain together. My second bowkill was unfortunately not as clean as my first. After I hit the ram he ran uphill and out of sight with his large buddy. After 14 hours of looking I finally spotted his companion feeding alone on a ridge. By following a path from where I last saw them together to his present location, I located my [dead] sheep. I was certain they wouldn't have separated unless my ram had died.

The Approach—from Above or Below?

One basic decision to make when hunting Dall sheep is whether to hunt them from above or below. Sometimes there are no options. In rough country a hunter often has no choice but to come at the sheep from below. When sheep are in canyons, the approach is from the rim. However, most of the time a hunter needs to carefully evaluate the situation.

One major consideration is sheep don't look up very often. Once a hunter is above sheep the odds shift dramatically in the hunter's favor. Dall sheep look downhill about ten times as much as they look uphill. They also instinctively go uphill to escape. On one hunt in the Alaska Range I was positioned uphill from a band of rams while my partner approached from below. He passed on the shot and then stood in plain view of the sheep. As they started to move uphill, I stood up. They then went to a small promontory 200 yards below me and stood watching for 20 minutes. They did not want to move laterally with me above them. When I finally started to move across the ridge, they quickly ran the opposite direction and climbed to safety. Sheep feel vulnerable when a predator is above them. They will sometimes even let a visible hunter move into gun range from above if their route to escape terrain is blocked.

The mountain gullies and surrounding rock walls also make it difficult for sheep to locate some sounds. The echoes from a gunshot bounce around in mountainous country so much sheep may inadvertently run toward a hunter that is above them. Hunters who approach sheep from above often get an opportunity for closer following shots because of this. Unfortunately, sheep **are** able to accurately pinpoint the source of falling rock and will locate a noisy hunter.

There is also a psychological advantage to being above our quarry. This is related to the obvious physical advantage. It is much easier to move down and across a hillside to cut off moving sheep than to climb and catch them.

There are also disadvantages to approaching sheep from above. Water is much scarcer on ridge tops and rocky peaks. Consequently, considerable time and energy are spent descending—and returning—just for water. Frequently, hunters on top cannot see sheep directly below them. Even during a downhill stalk, sheep are often hidden by the curve of the mountain. This lack of visibility

HUNTING STRATEGIES

from above also ruins stalks as hunters reach unseen cliffs and have to turn back. The time and effort necessary to get above sheep while remaining hidden is also prohibitive in some situations.

Before sheep are located, hunters usually travel along the bottom of drainages. This allows hunters to cover more ground with less effort while remaining hidden. With the use of quality optics hunters can even spot sheep while in plain view and not spook them. Hunters traveling up wide drainages where sheep have to scan a large expanse of country for predators aren't likely to spook sheep over a mile away. By using good optics, these same hunters can spot and evaluate sheep from even greater distances.

The ability to spot all the sheep on a mountainside increases as you move away from the mountain, but so does the time needed to stalk them. As you move closer, the mountain contours and terrain irregularities hide sheep from view. Close-up hunters are also more likely to get pinned down or spotted as sheep appear above them.

The optimal situation allows for a final approach from above. A sheep hunter in this position has the best chance for success. The details of the entire stalk will depend on the circumstances of each hunt.

Trojan Sheep

In many circumstances, Dall sheep cannot be approached without being seen. Their habitat often makes an approach from above physically impossible and gives them a complete view of any danger below. There is one way to hunt sheep successfully even in these situations—I call it the Trojan sheep strategy.

I imagine whites have been used as long as the white sheep have been hunted by man. I picked up one set of whites in the Wrangells that looked weathered enough to be from the '60s. Because Dall sheep are accustomed to dark-colored predators, they don't get too alarmed when they see white animals—a fact which hunters can use to their advantage.

I personally used a white suit for the first time in 1988. I was unable to close the final 150 yards and get within bow range of several rams without being seen. So, I just crawled out into the open in my white suit. I began pawing at the new snow and pretended to feed as the sheep were doing. To my amazement, it worked! The sheep did not spook! They ignored me and went on feeding. I was able to reach cover after 100 yards in the open and finish the stalk successfully. The poor visibility in the driving snow certainly helped me in this situation, but I have used whites successfully numerous times since then in very **unfavorable** conditions.

Along with white suits, hunters should also use white covers for their packs and weapons. The more precautions hunters take to complete this deceit, the more likely the ruse will work. Personally, the white suits I use are complete with a head and small horns. I make small horns to eliminate the remote chance of being shot and because I want to portray a **young** sheep. Adult sheep—even rams—are protective of their young. They also feel less threatened by young animals. While wearing a "sheep suit," rams within 100 yards of me have moved **closer**. They want a better look at this apparently handicapped sheep—I don't move very gracefully on my hands and knees.

Since Dall sheep are slightly yellow—not pure white—I look for yellowish coveralls or tint white ones slightly with a yellow dye.

Lambs tend to dissociate from their mothers (Geist, 1971) and yearlings often travel alone. So single, young sheep don't alarm adults. However, I have found that **two** hunters in white suits are much more convincing. I can remember two occasions when sheep traveled over one-half mile to join my partner and me while we portrayed sheep. One time a sheep came within **ten feet** for two minutes before deciding we were not good company.

The way hunters imitate sheep is important. First of all, **don't be vertical.** Hunters often use whites to cross open areas at long distances from sheep. This is an excellent use of whites **if** the hunters remain bent over. Try to present a horizontal line—like a sheep's back—at all times. This can be very tiring, but it is vital to consistent success. I have walked up to one-half mile while bent over as sheep scrutinized me to decide if I posed a threat. This is another instance when conditioning can favorably affect a sheep hunt.

A skilled Trojan sheep (hunter) will **look, act, and sound** like a young sheep. Sheep meander when feeding, they look frequently, bed facing downhill, and don't stare at each other. Remember sheep behavior as you imitate them. They will be suspicious of you. Show your backside frequently and don't travel directly at them. Don't give sheep any more time to scrutinize you than absolutely necessary.

Use cover to hide most of your body while in the open. Leave just a little bit of white showing so they can keep track of you and not get nervous.

Here the author demonstrates the strategic use of rocks to partially hide himself as a Trojan sheep.

One last ruse is to bleat like a lamb when you are close and the sheep are nervous. Dall lambs sound exactly like domestic sheep. They are **very** vocal in their first year. I have used this several times to save a stalk when sheep were about to spook. Be prudent about using this; overuse can ruin the stalk.

I don't recommend this as your primary hunting strategy, nor for every hunt. It's main purpose is for crossing open ground to reach more cover. The Trojan sheep strategy should be used **sparingly**. I once had a ram spook at 500 yards when he saw me portraying a sheep in the early morning. I think he had just been stalked by other hunters using whites and was aware of the danger. Or, perhaps he was suspicious because it was too early for sheep to be moving along a creek bottom. Regardless, it is not a guaranteed method.

I have also used whites to get by ewes without spooking them. Then, I continued stalking a ram which never saw me use the whites. I have also been spotted by rams in hunting clothes within 300 yards and immediately dropped and rolled out of sight. After donning my whites, I reappeared and continued a successful stalk. Whites may also give a hunter a few extra seconds at the moment of truth as sheep try to figure out what you are.

My most satisfying use of whites occurred just last season. The situation involved 20 rams and 80 ewes together in a long drainage. The only approach was in plain view of all of them. During the eight-hour stalk I passed within 40 yards of several sheep. I was even spotted in an upright position with my face exposed at the last moment by another ram. He was uncertain about the strange white sheep and didn't alarm the others. His hesitation gave me time to draw my bow and shoot his unexpecting companion.

The last item about Trojan sheep you need to understand is **don't ruin it.** Never stand up in whites at close range—particularly after you have shot a ram. Sheep are adaptive—they do learn quickly. On one guided hunt, after my hunter missed his first shot at a bedded ram, we had to stand up and run in order to monitor the scattering rams. We were able to locate and dispatch his sheep, but not before several others had seen us standing in our whites—one within 40 yards. This made them **very** distrustful of white animals. After several days they would still scatter when one of their own band suddenly appeared. I always plan to remove my suit immediately after a kill—**before** other rams spot me. Otherwise, sheep will become much more suspicious of Trojan sheep.

Bowhunting

Bowhunting for Dall sheep **is difficult**, but not impossible. Many bowhunters have been successful and a few have even bagged several sheep. Like bowhunting most big-game animals, the challenge increases when you need to get within bow range. Sheep hunters who use a bow should be prepared to work harder, endure more discomfort, and have more patience than gun hunters. But the accomplishment will be much more meaningful.

The differences in equipment for bowhunting sheep are few. Because of the layers of clothing often worn while sheep hunting, I recommend a long arm guard that covers the upper arm as well as the forearm—mine is white in case I am wearing a sheep suit at the time of the shot. I use a white sling in the same situation. This allows me to crawl on all fours and still carry my bow—which is camouflaged with white limb covers. I also carry a white chest protector to hold down the bulky clothing. I suggest compound-bow shooters carry tools to dismantle their bow after the hunt. It is much easier to travel down steep mountain faces and through alder jungles without a bow strapped on your pack. Another solution is to use a take-down recurve or longbow to make traveling easier. I have a take-down recurve for just that reason.

The rough conditions encountered while sheep hunting are also a consideration for bowhunters. Bowhunting equipment can be very fragile. Choose gear which will withstand rough treatment. My first bow hangs on the wall next to my first sheep as a reminder of how tough sheep hunting can be on equipment. Another bow of mine currently lies under a winter's worth of snow in my abandoned tent as testimony to the unpredictable weather in sheep habitat. Hopefully, it will still be functional when the spring thaw uncovers it and I retrieve it.

One major concern for bowhunters is being scented. Sheep hunts are often lengthy trips in remote country. Bathing regularly and wearing clean clothes every day isn't possible. But the winds are so erratic in the mountains that I have seldom been scented by sheep at over 200 yards. Because of this remote chance of being scented at long ranges, many gun hunters just ignore the wind. However, at bowhunting ranges the wind becomes significant—it has ruined more of my stalks than any other factor.

I have not found a reliable solution to this problem. I try to stay below ridgelines as much as possible at close range to avoid the contrasting air flows commonly found there. Since wind changes typically occur at mid-morning and late afternoon, these are risky times to be really close to sheep. Bowhunters should usually avoid these times when planning the final stalk. Bowhunters just have to **wait out** unfavorable wind conditions. I once waited on a very cold, windy ridgeline for five hours within 150 yards of several rams. The wind **finally** changed and I finished my stalk. Because bowhunters often need to wait longer for favorable conditions, they should plan on more bivouacs and carry appropriate gear.

Sheep **smell very strong**. They can often be scented **before** they are seen—particularly at bowhunting ranges. Also, their strong-smelling beds can help bowhunters. Rolling in a sheep bed helps mask human scent. This may give a bowhunter a few extra seconds when most needed.

The last strategy concerning scent is the use of the new odor reducing/masking products. I have just started testing them and I have no conclusive evidence yet. I think they may be useful, but I doubt they are a panacea to the scent issue.

Bowhunters should also consider ambushing sheep on trails and using decoys to push them when hunting with a partner. Indians used to ambush Dall sheep in Alaska by using their knowledge of daily or seasonal trail use. It is more difficult for hunters who only hunt for several days to pattern sheep, but **it is possible**. Sheep can also be maneuvered toward a concealed hunter. One bowhunting client of mine was able to circle above ten rams while I played a sheep and moved them toward him. We guessed right as to which way they would move when they became nervous about me. All ten sheep moved to within 30 yards of him—while staring at me. Unfortunately, the ram we wanted was on the far side of the group. They spooked before he could get a clear shot.

I've also tried to move sheep up to me by throwing rocks below them. However, the rams were too perceptive; they spotted the rocks in midair and looked up at the source.

One last consideration for bowhunters is sheep are light-skinned and (generally) not hard to kill. Dall sheep consistently go downhill after they have been hit. The rough country they live in usually prevents them from climbing with any kind of serious injury. And, because of their color and the open country, they can be visually-monitored after a hit.

Timing of the Hunt

The open season for Dall sheep in Alaska has traditionally been from August 10th to September 20th. There are several variables affected by when you choose to hunt.

In the early part of the season the weather is generally warmer and drier. August temperatures can be as high as 70 degrees in most sheep country, while a late September hunt may bring sub-zero weather to northern ranges. Early-season hunts tend to bring at least some sunny days, but September hunts are sometimes wet from start to finish. There is also more likelihood of "weather" days spent in camp during late-season hunts. The amount of daylight also changes noticeably between the start of the season and the last few days.

Seventy-degree temperatures are possible in the early part of sheep season in Alaska. Here the author sunbathes during a 1980 hunt.

    First-of-the-season hunters often get to hunt unalerted sheep. But, since most of the hunting pressure occurs in the first **two weeks** of sheep season, getting away from other hunters can be a problem. Late-season hunters have fewer rams to pursue, but fewer hunters to compete with. The stormy weather in the latter part of the season also tends to bring larger, reclusive rams from their hiding places. Wary rams are often more relaxed in bad weather and less alert because they don't expect as much hunting activity.

    Trophy quality and difficulty in caring for it are both affected by when you hunt. Sheep hair is generally short in the early part of the season. This can make a mount look terribly thin-necked and unappealing. Late-season sheep have much longer hair—which is more attractive to most hunters; although some individuals think large-necked rams look relatively small-horned. One sheep I took during a limited October hunt has three-inch-long hair and a massive neck. I love these features regardless of the reduction in apparent horn size. It is just a matter of personal preference.

The warmer early season also makes it more difficult to keep sheep meat and capes from spoiling—while the colder, late-season temperatures alleviate these concerns. Additionally, the colder temperatures kill most of the bugs which can bother hunters as well as infest sheep meat. The colder temperatures also cover the slopes with snow and ice. This makes climbing much more dangerous and limits possible stalking routes.

Transportation

Dall sheep are a very desirable trophy. Alaska's sheep populations are seeing ever-increasing hunting pressure. One result is most experienced sheep hunters use airplanes to get away from other hunters.

The remoteness of most sheep country requires the use of airplanes for transportation.

Many sheep hunters in Alaska have their own planes or have

HUNTING STRATEGIES

access to a plane. Even if they don't actually fly-in to hunt, most hunters scout their sheep area from the air before the hunt. It is difficult for walk-in hunters to compete with airplane hunters. The success rate for walk-in hunters is about one-third that of fly-in hunters. I once walked for three days to get to my "spot"—only to find a plane parked there.

Currently, most of the large rams taken are first spotted from the air. This is partly because accessible areas get too much pressure for big rams to grow large or to stay around once they do. The Kenai mountains are an excellent example of this. They have both good road access and a well-developed trail system through them—and the statistics reflect this. Sheep hunters there have the lowest success rate in the state and produce the smallest percentage of large rams. I took two rams out of three hunts in the Kenai Mountains in the late '70s, but this would be much more difficult under current conditions.

There are some permit areas reserved for walk-in hunters which **do** produce big sheep because of the limited pressure—so it is still possible. There are also areas accessible by off-road vehicle which don't receive as much pressure as more accessible sheep habitat. A personal acquaintance of mine just took a forty-inch, heavy-horned sheep by using a four-wheeler and then walking several miles to his sheep. I personally walked in on my first six successful sheep hunts. And I know an individual who typically rides a three-wheeler within a few hundred yards of his ram **every** year.

Airplane hunting is generally more expensive than walk-in hunting, but the differences in the time involved may make the actual cost lower. On one walk-in hunt I managed a return ride from a small airstrip rather than spend three more grueling days walking with a loaded pack. The $100 price was well worth the time savings. Unfortunately, the recent rash of small plane accidents has reminded sheep hunters of the greater risk involved when airplanes are used.

However Dall sheep are hunted, it is a fascinating experience. A successful Dall sheep hunter has many reasons to be proud of the accomplishment. However, there is still more to a hunt once the sheep is down. The trophy and the meat still have to be properly cared for to **finish** the hunt.

Chapter 9

CARE OF MEAT AND TROPHIES

The quality of the trophy and meat obtained from a sheep hunt is directly related to the care they receive **in the field**. Due to the extended nature of sheep hunts, hunters may have to take care of the meat and cape for a week or more. Weather conditions often make this a difficult task. Unequaled table fare and one of the finest trophies available are the rewards for successfully meeting this challenge.

The quality of both the trophy **and** meat can be significantly affected by how the sheep is killed. Just this past season I witnessed an almost-perfect example of the best way to dispatch a sheep. The bowhunter I was guiding made a perfect heart shot on an unsuspecting ram below us. The ram stopped feeding, but did not get excited. He took a few steps and seemed puzzled about what had happened. After thirty seconds or so, he simply passed out and rolled a few feet into the creek below him. The result was the tenderest and tastiest sheep I have ever eaten. We also had a clean, blood-free cape. Contrast this to a scenario where several shots are needed from a high-powered rifle. Perhaps the wounded animal is chased for several hours. Finally, he falls two thousand feet and dies. Of course the quality of both the trophy and meat would suffer greatly from such treatment. **Of far greater importance** is the unnecessary suffering the animal would endure in this situation. **We must strive to make a clean kill out of respect for the animal.** As responsible hunters, we **have** to consider the **possible results** before shooting!

G. Fred Asbell with his ram. This animal died very quickly—it was reflected in the excellent quality of the meat. 33 7/8 x 13. Score--144 2/8 P&Y.

Once the sheep has been recovered, the hair should **immediately** be cleaned of as much blood as possible. This takes only a few minutes and greatly improves the appearance of the mounted trophy. This will also improve field photos. If the sheep doesn't happen to fall into a convenient creek, the blood can be wiped off with water carried for just that purpose. Sheep hair is **hollow**. Any blood that **dries** inside the hair will never come out. By carefully and gently wiping the hair with a clean, damp cloth, some blood can be removed. This will also keep the hair moist until it can be soaked at a later time. However, you must consider the temperature and the time in the field **before** doing this. The hide—as well as the meat—will spoil if kept warm and moist for very long. How long depends on how warm and how moist they are. The decision whether to use water on the hair has to be made at the time of the kill.

Skinning

As soon as the animal has been cleaned and pictures taken, it should be skinned. Skinning cools the meat and hide. The quicker they cool, the better condition they will be in.

CARE OF MEAT AND TROPHIES

First, find the best spot available to skin the sheep. Sheep are small enough to move several feet to facilitate this. A grassy or mossy spot will produce cleaner results than a shale or soil surface. Moss also soaks up blood and can be stuffed into the nostrils and down the throat. This helps prevent more blood from leaking out during the skinning process. If moss is not available, tissue paper or other absorbent material can be used.

photo courtesy of Bob Kelez Sr.

The amount of time and effort spent caring for a sheep in the field influences the quality of the mount. Bob Kelez Sr.'s sheep was well-cared for—as reflected in this excellent mount. 40 7/8 x 14 2/8. Score—170 6/8 B&C.

Next, position the sheep with his head uphill and lying on his side. By keeping the head uphill, less blood will come out and it will flow away from the cape. At this point it helps to have knives, steel, saw, game bags, and plastic bags out of your pack and ready to use. This keeps both your pack and the trophy cleaner. I also locate and clean several large rocks or boulders (if available) to place the meat on as it is removed. Assuming you have decided whether you want a life-size or a shoulder mount, you are now ready to start skinning.

As one side of the animal is skinned, remove the exposed meat. The meat will stay cleaner this way and it can also be cooled quicker. Spread the meat out on the clean rocks you prepared and let the cool rock draw heat out of the meat as the air helps dry it. When the sheep has to be turned over, wrap the loose skin around the exposed parts to protect them until you can finish removing the meat.

Once the skin is removed, several options are open. It can be placed in a nearby stream to remove more blood. It can be spread out flesh side up to dry. Or, it can be rolled up and kept moist until it can be soaked later. The action taken depends on the condition of the hide, the weather, and the time left in the hunt.

Meat and Horns

Since sheep hunting commonly involves a long return hike, most hunters de-bone the meat to reduce weight. Removing the bones helps cool the meat faster and allows you to remove more excess fat and inedible tissue—which further reduces weight. The disadvantage of boning the meat is more surface area is exposed to collect dirt and moisture. Thus, more precautions will be needed to prevent spoilage. However, the reduction in weight is **well worth** these inconveniences.

The last item taken off the sheep is the horns. I carry a Knapp saw for this. It takes only a few minutes to cut the top of the skull off. Cut horizontally through the skull midway down the eye sockets, through the ear canals, and out the back. This reduces weight as much as possible and is the method preferred by taxidermists. If the return hike is going to be particularly long or difficult the skull plate can be split vertically. The horns will be much easier to pack this way. They will have to be put back together later to be mounted, but this will not affect their qualification for any record-keeping organization.

Quality Care

After leaving the kill site there are several procedures which will improve the quality of both the meat and trophy. When packing the meat or hide, keep it out of direct sunlight as much as possible. Placing both near the bottom of your pack will

CARE OF MEAT AND TROPHIES

accomplish this. Lay the meat and hide out at night under an overhanging rock or a tarp to cool and dry. Cover them up before the sun comes up, preferably with a light-colored object to reflect sunlight. If a watertight bag is available, both meat and hide can be immersed in a stream safely for days. All these procedures and any others that keep the meat and hide cool and dry will be beneficial.

If you are going to be out more than a few days after taking a sheep, it will need more attention. The skin should be salted and/or fleshed. Three pounds of fine salt is necessary for a sheep cape and about ten pounds for the entire skin. Rub the salt in carefully, particularly around the edges and face. Then roll it up, drain it, and re-salt. Alternatively, and my choice on most sheep hunts, is to turn the ears and nose, split the eyes and lips, and thoroughly flesh the hide. Any good taxidermist will gladly show you how to do this. The **purpose** of salting or fleshing the skin is to **dry** it. This prevents the hair from slipping (falling off). Either method will accomplish this. But, by knowing how to flesh properly, you can reduce weight in your pack. Additionally, by **both** fleshing and salting skins, they will almost never go bad in the field.

Preserving the Trophy

After returning from the field with your trophy, there is one more **very** important decision to be made. The taxidermist must be selected. Taxidermy is an art. Some practitioners are artists, many are not. It is not difficult to tell them apart. Considering the amount of time, effort, and money invested in a sheep hunt, taxidermy fees should not be of prime importance when choosing an artist. Choose a taxidermist that can capture the image of your trophy as you remember it.

After a taxidermist has finished a sheep mount, the hunter's responsibilities begin. Sheep hair is **extremely** brittle. Sheep mounts should be handled with the utmost care and as little as possible. Protect them by hanging them high and away from smoke. They are meant to be **seen and not touched.** Handle them by the horns and backboard if possible. If the hair must be touched, gently grasp the muzzle where the hair is more flexible or slide a clean hand onto the neck in the direction of the hair. Once a year, lightly vacuum the mount and wipe gently with a moist cloth. A sheep mount can last a **lifetime** if these precautions are taken.

A beautiful Dall ram mount by Larry McMurphy—owner of Woodland Taxidermy--Wasilla, Alaska. Larry lives in Alaska, mounts dozens of rams a year, and hunts them annually. Larry's artistry is evident in each mount. 42 x 13 1/2. Score—165 B&C.

The quality of field care is reflected in the quality of table fare as well as the trophy hanging on the wall. And the appearance of a trophy depends on the **continuing** care it receives. Plan ahead. Learn what is necessary and take what is needed for proper field care. Do all you can to preserve the priceless memories of the hunt. Both you and the animal deserve that respect.

Chapter 10

A TROPHY SHEEP

Dall sheep are one of the premier trophy animals in North America—perhaps in the world. They make beautiful mounts and draw admiring gazes in any trophy room. Experienced sheep hunters especially understand the time, effort, and skill invested in a sheep trophy. They are the most appreciative of a fellow hunter's accomplishment. However, the standards used to judge the quality of a sheep trophy vary greatly and are difficult to apply in the field.

Every hunter should decide what qualities **their** trophy sheep must have. Personally, I have always been looking for a ram with heavy horns that are broomed off evenly just at full-curl; I am **still** looking. A common standard many sheep hunters share is that of the "forty incher." Other experienced hunters are only interested in a "Book" ram. For most first-time sheep hunters **any legal ram** would be a trophy. Regardless of preferences, hunters have to **accurately** assess a sheep in the field to choose their trophy.

Legal rams

Currently, a legal ram in Alaska is full-curl. There are a few areas where any sheep is legal, but these limited areas are not of interest to most hunters. Hunting regulations in Alaska went from 3/4 curl in 1951 through 1978 to 7/8 curl from 1979 through 1983. In 1984 the present 4/4 minimum went into effect in many areas. It is now the standard throughout 95% of Alaska. The first guideline a sheep hunter needs to understand is what constitutes a **legal** ram.

The current Alaska State Hunting Regulations define full-curl as meaning:

(A) the tip of at least one horn has grown through 360 degrees of a circle described by the outer surface of the horn, as viewed from the side, **OR**

(B) the tips of both horns are broken, **OR**

(C) the ram is at least eight (8) years of age as determined by horn growth annuli.

There is some ambiguity about how to determine a full-curl horn using these regulations. The annuli are fairly easy to count and agree upon **if** the horns can be handled. I personally have only once been able to count the rings with 100% confidence in the field. If a ram is ten or eleven years old and not broomed (broken tips), an experienced hunter with good optics could reliably determine he was legal by the annuli method—most of the time. Sometimes the rings are just not distinguishable from other grooves in the horns. If a ram is **only** eight years old it is difficult to see all the rings and be 100% sure.

Rams with broken horn tips are usually identifiable, but sometimes worn tips look broken and vice versa. I recall one ram I studied for several minutes at thirty yards. Even with eight power binoculars I **still** couldn't tell if his horns were broomed or just worn. I settled for one of his companions whom I was sure was legal. Also, Dall rams do not break horn tips with nearly the frequency of Bighorn sheep, so, there are just not many broomed rams around. I am not certain I have even **seen** a Dall sheep in the field with both horns broomed. The 360 degrees criterion is most frequently used to determine legality.

My suggestions to help determine if a horn has grown through 360 degrees are to look from as many angles as possible and practice on mounted sheep. The *Alaska Hunting Regulations* booklet also has pictures and diagrams to help identify full-curl horns. The difficulty in identifying legal rams is another good reason to have high-quality optics when sheep hunting. I often spend hours glassing a ram within rifle range to determine if he is legal before shooting.

A TROPHY SHEEP

Record-Book Rams

Serious sheep hunters frequently spend years in pursuit of a "book" ram. After taking one or two full-curl sheep, a hunter often seeks a bigger challenge. Trying to take a sheep large enough to make the Boone and Crockett book is that challenge for many sheep hunters. This quest is usually unsuccessful—even serious sheep hunters might never see such a large ram in a **lifetime** of hunting. However, an understanding of how sheep are scored and recorded will help any hunter judge a trophy animal.

There are currently three record books which list Dall sheep:

Records of North American Big Game by the Boone and Crockett Club lists any legally-taken Dall sheep with a minimum score of 170 inches;

Bowhunting Big Game Records of North America by the Pope and Young Club lists Dall sheep with a minimum score of 120 inches legally taken with bow and arrow;

Records of North American Big Game by the Safari Club International lists legally-taken Dall sheep with a minimum score of 155 inches.

All three clubs score Dall sheep by first measuring the length of each horn to the nearest eighth inch. Next, the circumference of each base is measured—also to the nearest eighth. Then, using the longest horn length, three more circumferences of each horn are taken. One each at one-fourth, two-fourths, and three-fourths of that length away from the bases. From here on the scoring methods differ. The Safari Club simply totals these ten measurements for the final score. The other two clubs compare the circumferences of the two horns at the base and at each quarter and subtract any differences **before** totaling the ten measurements. The Safari Club also allows the horns to be scored **immediately**, but there is a 60-day drying period required by the other two clubs **before** scoring.

Field Judging

If a hunter is looking for a high-scoring ram, there are several generalities that will help judge sheep horns in the field. The first is that **mass** is more important than length. Obviously, eight circumferences can affect the score more than two length measurements. A quick look through either of the three record books will help illustrate the importance of mass to a high score. One excellent example is the pick-up listed in the Ninth Edition of the Boone and Crockett book in the number eleven spot. Although these Dall sheep horns are only 39 inches long, their impressive score of 182 points results from **incredible mass**; they have bases of 15 1/2 inches and third quarter circumferences of 10 1/8 inches. With this relative importance of length and mass in mind, there are some ways to estimate these in the field.

First, look at the overall shape of the horns. There are **parallel**, **diverging**, and **converging** shapes among Dall sheep horns. Parallel horns drop straight down from their highest point to their lowest point. I will refer to this as the basic shape. A converging set of horns angles in toward the jaw as the horns come down. Then the tips commonly flare out as the horns come back up toward the front. The number two sheep listed in Boone and Crockett's Ninth Edition is of this type, also referred to as Argali-type horns. Many hunters consider this type of horn configuration the most spectacular because

Alden B. Worachek II with his 40" ram from the Tok permit area. These diverging horns—also called flaring—make a dramatic mount. 40 4/8 x 13 4/8.

of the way the horns repeatedly change directions—creating an impressive shape. Diverging (also called flaring) horns angle out from the head continually from the bases to the tips. The current number one Dall sheep in Boone and Crockett is of this type. Notwithstanding the apparent contradiction to this next generality, converging horns tend to be longer when **compared** to diverging horns. This doesn't mean that diverging horns cannot be long, but when field judging horns, the parallel and especially converging types tend to be longer than they look. Diverging horns tend to be deceivingly short in length when compared to converging horns of apparently the same size. When trying to determine **absolute** size there are also some guidelines to consider.

The average length of a ram horn in Alaska when it reaches 360° is about 33 inches. Because of hunter selection, the average **harvested** 360° horn is about 36 inches. Of course sheep horns vary considerably in size, shape, and growth patterns. These variables

Bob Kelez Sr. with his 41" ram from the Chugach Mountains. These converging horns are typical of Chugach range sheep. Score—170 6/8 B&C.

are affected by habitat and genetic make-up. When estimating horn length it is necessary to look at how far they go up, back, down, and forward when viewed from the side and the front. Sheep horns vary in the amount they grow in each of these directions and any one can have a significant effect on horn length. Typically, experienced hunters look at how far a sheep's horns drop to determine length. The jaw is used as a reference point to decide how much drop the horns have. If the lower edge of the horns come down to the bottom of the jaw the horns are probably longer than average. Although

photo by Frederick L. Wedel JR.

Look how low a sheep's horns drop before turning upward when estimating length. These drop almost to the chin and are 42 1/2 inches long. Score—171 P&Y.

A TROPHY SHEEP

drop is the most significant of the four directions, the others are also important and can significantly affect length. Therefore, any variation from the average in how much horns go in any direction must be considered. The best way to get a good feel for the "average" is to look at as many sheep horns as possible. Looking at pictures, mounted trophies, and live sheep are all good practice.

Before estimating length, though, a hunter must try to establish the relative size of the animal. If the ram is an average ram with an average-size head you have a standard to go by, but there are **exceptions**. I once guided a hunter to a ram that had been spotted, judged, and unsuccessfully hunted by several serious sheep hunters. His size was estimated at up to 45 inches with a score in the high 170's. Upon taking the ram we discovered he had a normal-sized body, but a **short** face. His horns were 38 inches long and he barely scored 160 points. Because everyone had assumed his face was

A side view of Will Stewart's heavy-horned ram displays the relatively small "hole" outlined by the inner surface of the horns—consistent with this approach to judging mass. Score—170 B&C.

average, he was misjudged by almost everyone—myself included. If a ram can be compared to several others to establish his "averageness," judging horn length based on head size (or body size) is much more reliable.

Judging length is further complicated when mass has to be considered. One common method for judging mass is to look at "hole" size. When viewed from the side, the size of the gap inside the horn curl, coupled with a good estimate of horn length, is a good indicator of horn mass. The smaller the hole, the more mass the horns have and vice versa. The problem is that length and mass distort each other. Massive horns look shorter than they are and long horns look thinner than they are. So the "hole" method is useful only if you can first estimate length accurately.

When estimating the circumference of horn bases a hunter needs to view them from as many angles as possible. From the front, horns with heavy bases will look like they are growing over the side of the head as well as almost touching on top of the head. A smaller gap on top of the head obviously indicates larger bases. One indicator of base size from the side view is the ear. Heavy bases will tend to grow back to the ear while there will be a larger gap in front of the ear if the bases are small. This is a rough estimate at best. The "hole" method can also be used to get an idea of base size from a side view. It is a good practice to estimate both ways and compare the results.

Dall sheep horns tend to be triangular-shaped in cross section and less massive than bighorn sheep horns, which are more oval-shaped. Dall sheep, along with Stone sheep, are called **thin-horn** sheep because of their relatively slender horns. In general, the more oval-shaped a Dall sheep's horns are the more mass they have. Cross-sectional shape is best determined in the field by looking at the horn bases from above and behind the sheep. By looking at the inner surface of the horn bases, a hunter can distinguish triangular from oval-shaped horns by whether that surface is flat or rounded. The more the bases bulge inward toward each other on the top of the head the more mass they will have. Of course this is a **relative** factor. The **absolute** size of the front and side dimensions has to be taken into consideration first.

After estimating base size a hunter has to estimate how much mass the rest of the horn has. Since three more circumferences will be taken on each horn when they are scored, the amount of mass carried out toward the tip is crucial for a high score. Most Dall sheep horns get thin quickly after the bases. The farther out a set of horns carries it's mass the more it will score. Generally, hunters compare the rest of the horn with the bases to judge massiveness. There are several factors to consider when making this estimate.

A rear view of Will Stewart's ram. The bulging inner surfaces of the bases reveals their exceptional mass. Score—170 B&C.

First, broomed horns will look heavier than they are. The thick tips of broomed horns will make the entire horn **look** more massive than it really is. Darker horns also look heavier than they are. When viewed against the white coat of a Dall ram, dark horns stand out much better than lighter horns and, thus, **look** thicker. Long horns will look **thinner** than they really are. And horns that seem to be heavy at the second and third circumferences don't **necessarily** have heavy bases. Bases should be judged **independently** from the mass of the rest of the horn.

The best field-judging tip I can give you is to get as much **practice** as you can. Look at photographs of Dall sheep and mounted trophies. Walk around the trophies to get several views—then guess the score. Always try to estimate the score before someone tells you. I find that my first guess is usually the best. If I let someone else's opinion sway me, I commonly do worse at estimating score. Another axiom that seems to hold true is—you will know a big sheep when you see one. One look is all you usually need if the ram is indeed, exceptionally large. If you have to look several times to convince yourself, he is probably not that big.

Regardless how interested you are in the **score** of a Dall sheep's horns, knowing how to **compare** them will help you select your trophy in the field. Forty inches of length and fourteen inch bases are both exceptional measurements for a Dall sheep. Jack O'Connor only took one forty-inch Dall by the time he wrote "Sheep and Sheep Hunting." Only **three** Dall sheep in the current Pope and Young record book have bases of fourteen inches or more. These standards can certainly be challenging goals by themselves, or they can just be useful guidelines for field judging.

Any full-curl Dall ram is a trophy. It represents time, effort, and skill—besides making a magnificent addition to any trophy room. Every hunter has unique reasons for hunting sheep and unique requirements for a "trophy." Even hunters on fully-guided hunts should use their personal guidelines when selecting a trophy. Therefore, **every** sheep hunter needs some field-judging abilities.

Chapter 11

WHERE TO HUNT

Jack O'Connor wrote, "Most elementary is that it is exceedingly helpful to know where ram country is." His advice is timeless. As modern-day hunters, we will encounter ever-increasing competition for game animals. Sheep hunters who are consistently successful and occasionally have an opportunity at an exceptional ram will not only be skilled in the field, they must also be skilled researchers. Finding productive new hunting territory takes time, money, and perseverance.

Sheep hunters should keep in mind their particular strengths and weaknesses, as well as personal preferences, while searching for ram country. The rigors of the hunt, cost, time involved, type of hunting experience desired, style of horns, and the size of the trophy are all factors that need to be addressed by every hunter.

Nonresidents of Alaska must also consider one additional factor—a guide. Unless hunting with a relative within the second degree of kindred, nonresidents hunting Dall sheep in Alaska must hire a registered guide. Since any competent guide will have already selected an area to hunt, the area and the guide are basically chosen at the same time. Consequently, nonresidents should know the hunting variables associated with the different mountain ranges in Alaska. In this way, the guide/area they choose is much more likely to satisfy their unique requirements for a sheep hunt.

Information Sources

There are several sources of valuable information available to sheep hunters. The statistics produced by ADF&G are a primary source of information for serious hunters. This state agency compiles harvest reports by mountain range and even individual drainage. Biologists then determine success rates of residents and nonresidents, mean horn size, large ram production, and other statistics depending on funding available. Unfortunately, they don't have funds to keep track of all these parameters every year. However, the statistics they do have **are** available to the public. This information can obviously be extremely helpful when choosing a hunting area.

One factor you must be aware of, though, is the possessiveness of sheep hunters toward "their" hunting areas. This trait causes some hunters to report their sheep harvest inaccurately to protect their

Kelly Stevenson with a thick-horned Chugach ram. A serious sheep hunter, Kelly wouldn't even tell her husband the location of this ram before she killed it. 37 4/8 x 14. Score—160 B&C.

"hot spot." But, even though not all the statistics from ADF&G are complete or 100% accurate, they are still **very** useful.

Other excellent sources of information are the record books. A hunter interested in a large sheep can find where the largest heads have come from by examining the three record books. The same caution about misinformation also applies here, but there is still a considerable amount of good information in these books.

Another vital source of information for sheep hunters is topographic maps. The United States Geological Survey (USGS) has topographic maps of Alaska that are essential to planning and successfully completing a sheep hunt. The contour lines, geographic features, waterways, roads, etc. depicted on these maps provide hunters with invaluable information. Many sporting goods stores sell these maps, as do the USGS offices. A publication entitled *ALASKA Index to topographic and other MAP COVERAGE* is also available from the USGS. This publication is very useful when deciding which maps are needed for a particular hunt. The first thing I do to **my** maps is cover them with clear contact paper. This makes them waterproof and much more durable. To further protect them, I fold them and put them in Ziploc bags.

Of course, hunters themselves are one of the best sources of information. Experienced sheep hunters have first-hand knowledge of where the **big ones** are. Although some are very tight-lipped about location, almost all of them are willing to tell stories about the one that got away. With a few queries to other hunters, guides, air-taxi operators or just by keeping an ear open whenever hunting is discussed, the missing information can often be filled in. The mountain range can almost always be ascertained in this manner. Often, the **exact** location can be pinpointed by diligent research and a little luck.

FNAWS

The single best gathering of sheep hunters occurs at the annual national convention of FNAWS. This conservation organization is dedicated to protecting and enhancing the wild sheep populations of North America. During these conventions, hunters, guides, outfitters, artists, manufacturers, retailers, state wildlife departments, and authors gather to talk about sheep. Needless to say, there is an overwhelming amount of information available about sheep at these conventions. It is the next best thing to actually hunting sheep.

There are also local chapters of FNAWS. The Alaska Chapter has it's annual convention in Anchorage in late winter. Because of availability, the Alaska Chapter commonly displays many top Dall sheep heads. This display usually includes the number one ram taken by Harry L. Swank, Jr. This display of top Dall trophies is unmatched and should be seen by every serious sheep hunter.

Dall Sheep Management in Alaska

The most recent population estimate for Dall sheep in Alaska was 70,000 animals (Heimer, 1984). This population is dispersed among seven identifiable mountain masses in Alaska. They are the Brooks Range, the Tanana/Yukon Uplands, the Alaska Range (usually split into East and West portions for statistical purposes), the Talkeetna Mountains, the Wrangell Mountains, the Kenai Mountains, and the Chugach Mountains.

Within these mountain masses, sheep distribution is determined by climate. Because of heavy winter snowfall and insufficient wind to clear feeding areas, sheep do not live on the southern slopes of the Alaska, Chugach, or Kenai Mountains (Heimer, 1984).

The existing sheep populations are managed for viewing, maximum hunting opportunities, quality hunting, or trophy hunting. Approximately 70% of Alaska's sheep populations are available for hunting. The other 30% are available for viewing only (Heimer, 1984).

The first aspect to look at is management. Trophy areas usually have larger sheep and provide a higher-quality hunting experience. These areas also require a drawing permit to hunt. Areas managed for maximum hunting opportunities allow unlimited participation, but are often crowded and sometimes don't have any large rams. The **current** population and success rates of all areas, including permit areas, must also be considered as they vary **widely**.

Brooks Range

The Brooks Range has the highest population of sheep of any mountain mass in Alaska, but it is experiencing a current decline (Heimer, 1993). In the summer of 1992 I spent four days walking though a recently-closed portion of the Brooks and I saw numerous sheep carcasses. The biologists were still assessing the reason for the die-off when I related my observations to them. Also, due to National Parks, only 50% of the sheep in the Brooks Range are available to sport hunters. Nonetheless, the Brooks has a lot to offer a sheep hunter.

The physical challenge of the Brooks Range is moderate by Dall sheep standards. They are often described as rolling hills, which is a fair description when compared to most sheep habitat in Alaska. Of course, there are unclimbable portions of the Brooks Range. In **general**, though, this area would be a good choice for older or less-physical sheep hunters.

Carl Brent's bow-killed ram. This Brooks Range ram is the same age and length as the author's fourth Chugach ram (page 77), but this ram is heavier and has a higher Pope and Young score! 35 1/8 x 13. Score—147 3/8 P&Y.

Because of their location in Northern Alaska, the Brooks Range also requires more time and money to hunt. Conversely, this means they can provide a high quality hunt because of low hunting pressure.

The success rate of sheep hunters in the Brooks Range in the five seasons from 1987-1991 was 59%. This is a very good success rate for sheep hunters, but many hunters are critical of the relatively small size of Brooks Range rams. The common perception is—because of the short growing season and cold climate the sheep there grow slowly and don't get very large. The 1975 study by W. Heimer and A. Smith basically contradicts this perception. They found that Brooks Range rams were **fairly** average in size compared to rams in other ranges in Alaska. Although there are only two rams listed in the B&C record book which came from the Brooks Range, two top-ten rams were taken in the Brooks just a couple years ago. Personally, I believe as hunters continue to explore the Brooks Range, they will find the potential for large rams is greater than we thought.

Alaska Range

The Alaska Range has almost as many sheep available for hunting as the Brooks Range (12,000 versus 15,000 respectively, ADF&G, 1984). The Alaska Range also has a lot of guiding activity which pushes up the success rate—guided hunters tend to have a much higher success rate than non-guided hunters. The success rate for the five-year period from 1987-1991 in the Alaska Range East was 38%. In the Alaska Range West it was 61%. The difference is mostly due to the higher percentage of guided hunts in the western portion of this range.

These mountains tend to be average Alaska sheep mountains. They are steep, rugged, and have plenty of glaciers. They also produce a fair number of record-book sheep.

There are two trophy areas in the Alaska Range East. They both produce some large sheep each year and hunters are limited by drawing permits. One of the areas even restricts motorized travel by hunters to the last half of the season. Both areas provide a high-quality hunting experience for those who are lucky enough to draw permits.

However, drawing permits don't always guarantee there are large sheep in the area. I was a lucky permit winner in one of these trophy areas several years ago. I climbed a lot of mountains searching for a large ram. I did see over fifty legal rams, but I never saw an exceptional ram. Eleven days of hard walking taught me that finding a record-book sheep takes more than just winning a special permit.

Tanana/Yukon Uplands

The Tanana/Yukon Uplands is a small area with a low-density sheep population. Because the access is difficult it is mostly hunted by locals. The success rate for the recent five-year period was low—27%. It has produced some large sheep—I've seen a mount of a 165-point ram taken there several years ago. I can attest to the difficulty of getting to this hunting area as well as the scarcity of sheep. On a recent hunting trip to these mountains, my partner and I traveled three days on foot to get to our hunting area, but found no sheep. I've heard the same story from other hunters and, like them, I vowed to return someday. These sheep have a unique characteristic I am looking for.

Kenai Mountains

The Kenai Mountains are very accessible. Due in part to this fact, the recent five-year success rate was the lowest in the state—17%. This range also tends to produce small sheep because most of the legal sheep are taken each year. The population has also been declining lately. One reason for this is the relatively-easy terrain found there. I remember one particular mountain in the Kenai Range that I hunted three years running in the late seventies. Whenever the sheep were pushed, they would always run to the same cliff face for safety. The trouble was, the face was navigable and also approachable from above. The rams were easy pickin's for any sheep hunter worth the label. The two sheep I chose to shoot two years apart were shot within 100 yards of each other. And neither knew anything was up until it was too late.

The **massive** number Eleven B&C ram I referred to in the trophy chapter came from the Kenai mountains. This area does have the ability to produce big sheep. As recently as last year a fellow hunter took a nice ram that almost made the B&C record book. Also, because of their location, hunting the Kenai Range can be one of the least expensive sheep hunts in Alaska.

Talkeetna Mountains

The Talkeetna mountains are another fairly small range that is easily accessible. Because of this, they get a lot of hunting pressure. The success rate has been only 25% for the most recent five-year period. Due partly to heavy hunting pressure, the rams tend to be small in the Talkeetnas. I discovered this for myself several years ago. I walked ten miles cross-country to escape other hunters **and** find a big ram. I was disappointed to find the biggest ram in the area was only 27 inches. He had pencil thin horns—despite sporting a full curl.

The Talkeetnas also tend to provide easy traveling. One client of mine last year shot his sheep at less than 3,000 ft. in a creek bottom. I've even had a client who shot a forty-incher in a brushy canyon in this area. However, like most sheep habitat in Alaska, the Talkeetnas also have some rough areas full of glaciers.

There are a few large sheep produced in the Talkeetnas. The problem is the same as in other accessible areas—the rams get cropped before growing large horns. A couple I know recently had their eyes on a large ram in the Talkeetnas, but were beaten to the punch on opening morning by local meat hunters. The locals didn't know what they had and simply took the meat and horns and cut

The Talkeetnas are capable of producing large rams like this one taken by Dr. David Mosal. As his guide, the author is equally happy with this 40" ram. 40 x 14 2/8. Score—165 B&C.

the cape at the neck. They were not impressed by the 39.5-inch horns that would have thrilled most serious sheep hunters.

Chugach Mountains

The Chugach Mountains are geologically young and, therefore, very rough mountains. There are many areas that are impassable and quite a few large, glaciated areas. Because there is a lot of escape cover for sheep, many rams survive to reach their full size potential. This is one reason the Chugach produces so many large rams.

The success rate was only 31% for the five seasons from 1987-1991, partly due to the numerous trophy hunters that hunt the Chugach. Since trophy hunters are looking for an exceptional ram, they often pass up legal rams and go home empty-handed, thus lowering the success rate.

The latest population estimate for the Chugach Mountains was 5,000. There are many areas accessible to walk-in hunters. One permit area is **restricted** to walk-in or horseback hunters. Because the Chugach is known for large sheep and it is close to Alaska's population center, the sheep are constantly being scouted from the air. Seldom does a good flying day go by during the hunting season without scouting flights throughout this range. As a result, large sheep there often learn to disappear during the season. Some live to an old age.

Even the walk-in permit areas of the Chugach get a lot of pressure. The only time I was able to draw a permit there, I saw seven hunters in seven days. This is high for a sheep hunt.

One of the open areas in the Chugach I hunt regularly has a very rough section that is almost impassable. Regardless of the risk, there are always hunters trying to scale this "wall" because of the large sheep found there. Such is the compelling power of sheep hunting for those of us hooked on it.

Wrangell Mountains

The Wrangell Mountains have a large sheep population and a high success rate. The latest figures put the population at 12,000 and the success rate at 52%. Guiding activity is high in the Wrangells—one reason for the good success rate.

The Wrangells have also produced a lot of record-book rams.

The one area in Alaska identified by Heimer and Smith (1975) as having the greatest potential to produce large Dall rams is in the eastern portion of the Wrangells. I can personally attest to the presence of large rams in this area. A few years ago I spent two weeks in July photographing sheep there. During that time I watched some of the biggest sheep I will probably ever see on the hoof. Unfortunately, this area is now a National Park and closed to sport hunting. Consequently, only a few subsistence hunters are allowed to hunt these trophy animals.

Much of the Wrangells are rough, glaciated mountains with difficult access—reachable by airplane only. However, there is one area restricted to walk-in hunters for those who want to avoid airplane hunters, but it does receive a lot of pressure.

What do YOU Want?

In general, there are several approaches to avoiding the crowds and realizing a quality sheep-hunting experience. One good way is to win a drawing permit which limits hunters. You can also pay to fly to remote country or pay a guide who can take you to little-used hunting grounds. If you have more energy than money, you can just pick an area which is only accessible by several days of walking. Definitely try to avoid most of the Kenai, Talkeetna, Chugach, and western Wrangells because of their proximity to Alaska's population centers. These areas always seem to have heavy hunting pressure.

If you are looking for an area with lots of sheep, most permit hunts also satisfy this criterion. Otherwise, go to the Brooks Range or the Wrangells.

An inexpensive hunt can be experienced in parts of any of the mountain ranges for resident hunters willing to out-walk their competition. Study topographic maps for areas reachable by the road system and query other hunters who have experience there to find the best routes to travel. Be forewarned that walk-in hunters generally have a low success rate—often below 10%.

For those hunters requiring a physically less-demanding hunt, the Talkeetnas, Brooks Range, or Kenai mountains are common choices. However, with just a little research, specific areas can be identified in all the ranges for an easier hunt. Look at the closeness of topographic lines and the amount of glaciation present when studying maps. These are good clues to the difficulty of the terrain.

When size is the main criterion for deciding where to hunt, there are numerous factors to consider. Permit areas often hold a few large sheep due to excess legal rams which are allowed to grow old. The Wrangells and the Chugach mountains have historically produced the most record-book rams. This is in part because of the extremely rough terrain and high incidence of glaciation in these ranges. Both provide escape cover—allowing rams to avoid hunters and grow large.

Another strategy is to walk a long distance into country only approachable on foot and look for the wise, old ram who thought he was safe from hunters. Or, you can wait until the end of the season and hope the approaching winter weather will prematurely bring a large ram out to more accessible terrain.

The largest ram I have personally taken was due to early winter weather. Fortunately for me, he let down his guard when a dense snowstorm cut visibility to 100 yards. He hadn't learned that serious sheep hunters are undaunted by life-threatening weather—as we should be. I arrowed him on a valley floor at 17 yards—in terrain he wouldn't frequent during normal weather conditions. This Dall ram is currently #1 in the Pope and Young record book. Sometimes a little extra effort is needed to take a trophy ram.

Of course you will need time and patience if a large ram is your goal. Most of the large sheep taken currently are what I call "airplane sheep." Trophy hunters commonly spend countless hours scouting with a super cub to locate an exceptional Dall ram. Hundreds of hours of flying time representing thousands of dollars goes into the taking of many record-book rams. Often the hunt is anticlimactic because sheep are not really that hard to take for experienced **rifle** hunters. In contrast, this past season a fellow hunter took a real trophy ram on the second-to-last day of the season by just putting in his time in the field. He had no idea the ram existed until he spotted him on that memory-making morning. He hunts hard and I consider him a real sheep hunter. It somehow seems more meaningful to take this type of trophy rather than an "airplane sheep."

The occurrence of a truly large ram in any population is rare at best. In his paper on *Alternative Rutting Strategies in Mountain Sheep: Management Applications*, Wayne Heimer (1990) goes into a detailed explanation of the conditions he feels are necessary for a super ram to appear. This paper is fascinating reading. It is recommended reading for anyone looking for an exceptional ram.

Todd Ripple's 40" ram. This sheep went unnoticed in a heavily-hunted area. It pays to scout very carefully. 40 x 13 2/8. Score—159 B&C.

The choice of where to hunt depends on the unique set of goals of each hunter. I have given many generalities to convey as much information as possible. Individual research is necessary to find the appropriate combination of factors for **your** sheep hunt. Your trophy is waiting—you just have to **find** it.

Chapter 12

REFLECTIONS

The beauty of sheep country is unsurpassed. The perspective from high on a mountain peak is spectacular. It always elicits a moment of inner satisfaction and wonder as the endless miles of ridges, valleys, crags, and glaciers lay seemingly within our grasp. From these lofty heights, the image of a dramatic sunset or a threatening cloud formation as a storm develops can create an unforgettable memory—one that stays with us for the rest of our lives.

Sheep hunting scenery is unmatched. This eastern portion of the Wrangell mountains is also well-known for large sheep.

Memories are the real trophies of a sheep hunt. Although a mounted sheep trophy can be a tribute to our hunting prowess, it primarily serves as a stimulus to bring back a flood of memories associated with the hunt. They also remind us of how much there is to learn about sheep hunting.

The author took this 3/4-curl ram while hunting with his brother, Randy (pictured). The author was hooked on sheep hunting after this 1978 hunt—even though hypothermia almost killed both Randy and Tony. 28 x 13 6/8.

I consider myself an experienced, competent sheep hunter—with a lot to learn. Each conversation with another sheep hunter—even a novice—can teach me something—if I listen. Part of the challenge of every sheep hunt is to become a better hunter. As I pursue this challenge, I share a responsibility with every other hunter.

We must help preserve the opportunity to hunt—for ourselves and those who follow. First, we must be responsible, ethical hunters and set high standards for ourselves as a model to others. Second, we all must get involved in the battle to preserve our hunting rights. They will surely disappear if we don't!

REFLECTIONS

Wild game populations are a renewable natural resource we should manage and harvest wisely. Effective game management utilizes hunting as a vital management tool to harvest surplus animals. We are part of the balance of nature. That we should hunt animals for food is empirical—down to our soft-soled feet. This adaptation allows us to quietly sneak up on our prey—GOOD SNEAKIN' TO YOU...

The vast country Dall sheep inhabit is beautiful and forever drawing us back..

APPENDIX

OFFICIAL SCORING SYSTEM FOR NORTH AMERICAN BIG GAME TROPHIES

Records of North American
Big Game

BOONE AND CROCKETT CLUB

Old Milwaukee Depot
250 Station Drive
Missoula, MT 59801

Minimum Score:	Awards	All-time	SHEEP	Kind of Sheep: _____
bighorn	175	180		
desert	165	168		
Dall's	160	170		
Stone's	165	170		

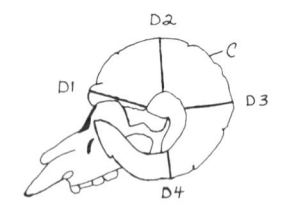

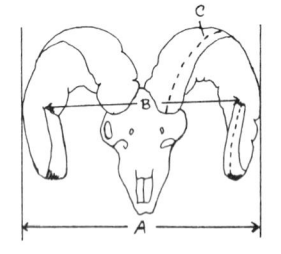

SEE OTHER SIDE FOR INSTRUCTIONS			Column 1	Column 2	Column 3
A. Greatest Spread (Is Often Tip to Tip Spread)			Right Horn	Left Horn	Difference
B. Tip to Tip Spread					
C. Length of Horn					✗
D-1. Circumference of Base					
D-2. Circumference at First Quarter					
D-3. Circumference at Second Quarter					
D-4. Circumference at Third Quarter					
	TOTALS				
ADD	Column 1		Exact Locality Where Killed:		
	Column 2		Date Killed: By Whom Killed:		
SUBTOTAL			Present Owner:		
SUBTRACT Column 3			Owner's Address:		
FINAL SCORE			Guide's Name and Address:		
			Remarks: (Mention Any Abnormalities or Unique Qualities)		

I certify that I have measured this trophy on _____ 19 _____
at (address) _____ City _____ State _____
and that these measurements and data are, to the best of my knowledge and belief, made in accordance with the instructions given.

Witness: _____ Signature: _____

B&C OFFICIAL MEASURER

I.D. Number

APPENDIX

INSTRUCTIONS FOR MEASURING SHEEP

All measurements must be made with a 1/4-inch wide flexible steel tape to the nearest one-eighth of an inch. Wherever it is necessary to change direction of measurement, mark a control point and swing tape at this point. Enter fractional figures in eights, without reduction. Official measurements cannot be taken until horns have air dried for at least 60 days after the animal was killed.

A. Greatest Spread is measured between perpendiculars at a right angle to the center line of the skull.

B. Tip to Tip Spread is measured between tips of horns.

C. Length of Horn is measured from the lowest point in front on outer curve to a point in line with tip. Do not press tape into depressions. The low point of the outer curve of the horn is considered to be the low point of the frontal portion of the horn, situated above and slightly medial to the eye socket (not the outside edge). Use a straight edge, perpendicular to horn axis, to end measurement on "broomed" horns.

D-1. Circumference of Base is measured at a right angle to axis of horn. Do not follow irregular edge of horn; the line of measurement must be entirely on horn material, not the jagged edge often noted.

D-2-3-4. Divide measurement C of longer horn by four. Starting at base, mark both horns at these quarters (even though the other horn is shorter) and measure circumferences at these marks, with measurements taken at right angles to horn axis.

FAIR CHASE STATEMENT FOR ALL HUNTER-TAKEN TROPHIES

FAIR CHASE, as defined by the Boone and Crockett Club, is the ethical, sportsmanlike and lawful pursuit and taking of any free-ranging wild game animal in a manner that does not give the hunter an improper or unfair advantage over such game animals.
Use of any of the following methods in the taking of game shall be deemed **UNFAIR CHASE** and unsportsmanlike:

I. Spotting or herding game from the air, followed by landing in its vicinity for the purpose of pursuit and shooting;

II. Herding, pursuing, or shooting game from any motorboat or motor vehicle;

III. Use of electronic devices for attracting, locating, or observing game, or for guiding the hunter to such game;

IV. Hunting game confined by artificial barriers, including escape-proof fenced enclosures, or hunting game transplanted solely for the purpose of commercial shooting;

V. Taking of game in a manner not in full compliance with the game laws or regulations of the federal government or of any state, province, territory, or tribal council on reservations or tribal lands;

VI. Or as may otherwise be deemed unfair or unsportsmanlike by the Executive Committee of the Boone and Crockett Club.

I certify that the trophy scored on this chart was taken in **FAIR CHASE** as defined above by the Boone and Crockett Club. In signing this statement, I understand that if this entry is found to be fraudulent, it will not be accepted into the Awards program and all of my prior entries are subject to deletion from future editions of *Records of North American Big Game* and future entries may not be accepted.

Date: _____ Signature of Hunter: _____
(Have signature notarized by a Notary Public.)

Copyright © 1993 by Boone and Crockett Club
(Reproduction strictly forbidden without express, written consent)

APPENDIX

SAFARI CLUB INTERNATIONAL
Method 11 Entry Form

For wild sheep, bharal, aoudad and eastern tur.

Diagram labels:
- Length of Horn
- C-1 (Circumference of Horn at 1st Quarter)
- B (Circumference of Horn at Base)
- C-2 (Circumference of Horn at 2nd Quarter)
- B (Circumference of Horn at Base)
- Length of Horn
- C-3 (Circumference of Horn at 3rd Quarter)

Hunter _____ (HOW YOU WANT YOUR NAME TO APPEAR IN THE RECORD BOOK)

Address _____

CITY _____ STATE _____ ZIP _____ COUNTRY _____

Ph. ( ) _____ HOME ( ) _____ BUSINESS ( ) _____ FAX

"I certify that, to the best of my knowledge, I took this animal without violating the wildlife laws or ethical hunting practices of the country or province in which I hunted. I also certify that, to the best of my knowledge, the laws of my country have not been violated by my taking or importing this animal."

Signature _____

TROPHY RECORDS OFFICE
- Species No. _____
- Record No. _____
- Batch No. _____
- Approved / Rejected by _____
- Date _____
- Medallion/Year _____ / _____
- Status: P ___ I ___ A ___ H ___ B ___ Top Ten ___

FOR ACCOUNTING USE ONLY

Animal _____

Remeasurement? ❏ Yes ❏ No Former Score _____ Record No. _____ (IF KNOWN)

Date Taken _____ MONTH _____ DAY _____ YEAR Free-ranging ❏ Yes ❏ No

❏ Rifle ❏ Handgun ❏ Muzzleloader ❏ Bow ❏ Crossbow ❏ Picked Up

Place taken _____ COUNTRY _____ STATE or PROVINCE

Locality _____

Game Management Unit if Applicable _____

Guide _____ Hunting Co. _____

I. Length of Horn   L _____ /8   R _____ /8

II. Circumference of Horn

B (At base)   L _____ /8   R _____ /8
C-1 (At 1st quarter) L _____ /8   R _____ /8
C-2 (At 2nd quarter) L _____ /8   R _____ /8
C-3 (At 3rd quarter) L _____ /8   R _____ /8

V. TOTAL SCORE   _____ /8

Official Measurer _____

Date Measured _____ MONTH _____ DAY _____ YEAR   Certified ❏ Yes ❏ No

Signature of Measurer _____

RECORD BOOK FEE $25.
MEDALLION FEE $30.

Submit to: Safari Club International, 4800 W. Gates Pass Rd., Tucson, AZ 85745 USA.
All entries must be accompanied by entry fees and a photograph of the trophy.
Please clearly label back of photo with name of hunter, name and score of animal, and date taken.
❏ 1 Photo included ❏ For animals with branched antlers: enough photos so that all tines can be clearly seen.
Checks on U.S. banks only. Credit cards preferred. Entry fees are valid for 12 months from date of form located in lower left hand corner.
We Accept: ❏ MC ❏ Visa ❏ AMX   _____ EXPIRATION DATE   _____ CARD NUMBER

1/94

COPYRIGHT © SAFARI CLUB INTERNATIONAL

APPENDIX

REFERENCES

Carper, Jean. Jean Carper's Total Nutrition Guide. New York: Bantam Book, Inc., 1987.

Geist, Valerius. Mountain Sheep and Man in the Northern Wilds. New York: Cornell University, 1975.

——. Mountain Sheep—A Study in Behavior and Evolution. Chicago: University of Chicago Press, 1971.

Harkness, David, comp., ADF&G Harvest Summaries. Alaska: ADF&G, 1991.

Heimer, Wayne E. Alternative Rutting Strategies in Mountain Sheep: Management Applications, from Proceedings of the Seventh Biennial Symposium of the Northern Wild Sheep and Goat Council. Washington: Northern Wild Sheep and Goat Council, 1990.

——. Toward a Working Hypothesis for Mountain Sheep Management, from Proc. of the Sixth Bien. Symp. of the NWSGC. Alberta: NWSGC., 1988.

——. Population Status and Management of Dall Sheep in Alaska, from Proc. of the NWSGC. Manfred Hoefs, ed. Alaska: NWSGC., 1985.

——. The Dall Sheep in Alaska, ADF&G Wildlife Notebook Series. Alaska: ADF&G, 1984.

Heimer, Wayne E., and Smith, Arthur C. III. Ram Horn Growth and Population Quality—Their Significance to Dall Sheep Management in Alaska. Alaska: ADF&G, 1975.

Heimer, Wayne E., and Watson, Sarah M. Report on Maximizing Ram Harvest, from Proc. of NWSGC. Montana: NWSGC., 1986.

O'Connor, Jack. Sheep and Sheep Hunting. New York: Winchester Press, 1974.

Oregon Freeze Dry, Inc. Nutritional Content—Mountain House Foods in Foil Packages. Oregon: Oregon Freeze Dry, 1989.

Stapleton, Lance. A Reference Manual for Hunting North American Wild Sheep. Florida: The Hunting Report, 1991.

HORN & ANTLER CARVINGS
by TONY RUSS

Native Alaskan, guide, writer, and bowhunter

Creations that display the beauty and majesty of Alaska **and** her wildlife, captured in materials from the animals themselves.

From glass-topped antler tables for the den to ram horn mantle pieces -- my design or yours -- your horns or my stock.

Call or write:

Tony Russ
574 Sarahs Way
Wasilla, AK 99654
(907) 376-6474

APPENDIX

GEAR CHECKLIST

Sleeping
tent
sleeping bag
sleeping pad
Cooking/Eating
stove & spare parts
fuel
pot
cup
spoon
water bottles—2
Clothing
boots
4-8 pr socks
shorts
long underwear
pants
shirt
coat
hat
gloves
raingear
bandanna or neck gaiter
Hunting
pack w/extra pins & rings
weapon
ammo
accessories
license & tags
spotting scope
binoculars
camera & film
game bags
plastic bags
sheath knife
pocket knife
sharpening tool
salt for cape

First-Aid
prescription medicine
Bandaids/Moleskin
anti-diarrhetic
antibiotic
Other
climbing pole
20' twine
20' light-weight rope
map
flashlight w\new batteries
toothbrush/paste/floss
waterproof matches
lighter
firestarters
small piece soap
toilet paper
sewing needle/buttons
safety pins
suncreen
lip balm
compass w\mirror
Food . . .

159